T0389087

Hazmatology
The Science of Hazardous Materials

Hazmatology: The Science of Hazardous Materials,
Five-Volume Set
9781138316072

Volume One - Chronicles of Incidents and Response
9781138316096

Volume Two - Standard of Care and Hazmat Planning
9781138316768

Volume Three - Applied Chemistry and Physics
9781138316522

Volume Four - Common Sense Emergency Response
9781138316782

Volume Five - Hazmat Team Spotlight
9781138316812

Common Sense Emergency Response

Robert A. Burke

CRC Press
Taylor & Francis Group
Boca Raton London New York

CRC Press is an imprint of the
Taylor & Francis Group, an **informa** business

CRC Press
Taylor & Francis Group
6000 Broken Sound Parkway NW, Suite 300
Boca Raton, FL 33487-2742

Printed on acid-free paper

International Standard Book Number-13: 978-1-138-31678-2 (Hardback)

Visit the Taylor & Francis Web site at
http://www.taylorandfrancis.com

and the CRC Press Web site at
http://www.crcpress.com

Dedication

Volume Four

James W. Covington

On February 24, 1978, following the Waverly, TN derailment and explosion, for which the Memphis Fire Department provided mutual aid, Memphis began looking at ideas for the formation of a hazardous materials response team in Memphis. Captain James Covington, Charles Smith, and Chief Adelman, Chief of Training, spearheaded the Memphis hazmat team formation in the Spring of 1978. When he retired from Memphis Fire Department, Jim went on to become a hazmat instructor at the National Fire Academy, where he taught drawing from his

Memphis hazmat experience and established himself as a national leader in the instruction of hazmat. That is where I met Jim, and we became friends after I took one of his hazmat tactics classes. His down-to-earth and knowledgeable style of teaching made his classes interesting and informative. Jim has since passed away, but his mark on hazardous materials response will live on.

Contents

Preface

When we are being called to the scene of a hazardous materials incident, it is because the situation is out of control. Our job is to stabilize the incident with a systematic approach based upon science and risk management. The first and most important task for emergency responders is to be able to determine if hazardous materials are present. If we cannot do that, nothing else matters. Once it is determined that a hazardous material is present, responders must consider all of the actors that are present. Actors include the hazardous material, container, physical state, nonhazardous materials, weather, and environment.

Following determination of the hazards, it is necessary to confirm who and what is at risk or in harm's way. What is at risk if we do nothing? Can we make the situation better than it would have been if we hadn't been there at all? What we decide to do should be based upon the risk to the public, responders, property, and environment, and not just because we can. Following the determination of hazard and risk and the decision to do something, we need to determine how to protect ourselves and the public.

Protection includes moving or sheltering the public, moving responders, utilizing personnel protective equipment, or letting the incident take its course. Protection requires proper numbers of personnel and the proper equipment to carry it out. Safety of the public and response personnel is the number one priority.

Once the safety and protection issues are addressed and the decision is made to address the hazards of the incident, tactical options need to be identified and implemented. Tactical options are based upon the overall incident goals and stabilization of the incident scene. Clean-up is not a responsibility of emergency responders. There are clean up companies either procured by the spiller or on a list at the state Department of Environment to conduct clean up operations.

Options for stabilization can vary widely depending on the scope and size of the incident and hazards involved. Science and risk analysis play an important role in the successful outcome of the stabilization portion of the response. Following incident stabilization, the response portion of the incident operations can be scaled back. During this period, restocking and documentation occur. Critiques are generally conducted formally some time after the conclusion of the incident. Some documentation needs to be done while all is fresh in the minds of response personnel to be utilized at the formal critique. This portion of the incident, while not as exciting as the tactical phase, is very important to do thoroughly.

Acknowledgements

I thank the many fire departments and members across the United States and Canada that I have visited and became friends with during my visits to their departments over the years. I also thank the firefighters from classes I have attended as a student and taught for the National Fire Academy, Maryland Fire and Rescue Institute and Community College of Baltimore County since 1988. Learning is a two-way street, and I have learned much from the students as well. I thank the many friends I have met during the 40 plus years in the fire, EMS, hazardous materials and emergency management fields. There are those who I have not seen for a while; some are no longer with us, but once a friend, always a friend.

I express my thanks to *Firehouse Magazine* for allowing me to write stories about hazardous materials for 33 years and counting. During those years, I have had the pleasure of writing under every editor of the magazine including founder Dennis Smith who gave me the chance to be published for the first time. I also thank Firehouse editors, Janet Kimmerly, Barbara Dunleavy, Jeff Barrington, Harvey Eisner, Tim Sendelbach and Peter Mathews for their support over the years. When I read my first copy of *Firehouse Magazine* in the late 1970s, I was hooked. My dream was to someday go to Baltimore to attend a Firehouse Expo. Never did I dream I would not only attend an expo but teach at numerous expos, write for the magazine and in 2018 be inducted into the Firehouse Hall of Fame. To be placed in a fraternity with sixteen of the people who had an enormous impact on the fire service and who I looked up to my entire career was very humbling.

Several people have been my mentors and have impacted my life and career. When I worked with the State Fire Marshall of Nebraska, Wally Barnett allowed me to accomplish things in the State Fire Marshal's Office

Brent Boydston, Chief Bentonville, AR Fire Department.

that I otherwise would not have. Because of his ability to let his employees reach their potential, I was able to write for *Firehouse Magazine*, become a contributing editor, teach for the National Fire Academy and other things too numerous to mention. He was proud when I gave him a copy of my first book. I owe much of my success in the fire service to the opportunities Wally gave me. Jan Kuczma and Chris Waters at the National Fire Academy have been mentors to me over the years. Ron Gore, retired Captain from the Jacksonville, FL Fire Department and Owner of Safety Systems, has had a large impact on my life and career. The Jacksonville Hazmat Team was the first emergency services Hazmat Team in the United States. Ron Gore is the God Father of Hazmat response in the United States.

Former student of mine and current Chief of the Bentonville, AR Fire Department Brent Boydston has been a great friend to me and my family over the years. Rudy Rinas, Gene Ryan and John Eversole of the Chicago Fire Department have been fellow classmates and students. Mike Roeshman and Bill Doty of the Philadelphia Fire Department both former students and retired as Hazmat Chief Officers have remained friends. I used to ride with Bill and together we had some great adventures. Mike showed me Philadelphia historical areas, like the spot where Ben Franklin flew his kite and his post office, which is so obscure today in downtown Philadelphia. I also stood on the spot where Rocky stood at the top of the steps in the movie. These adventures enjoyed in Philadelphia would not have happened without Bill and Mike.

Mike Roeshman Retired Hazmat Chief Philadelphia Fire Department.

Just outside of Philadelphia in Delaware County, Tom Micozzie, Hazmat Coordinator for Delaware County, was also a former student and a great friend. We had many adventures together, and I will never forget his introduction to me of the Galati at Rita's Italian Ice! Rita's Italian Ice was started by a retired Philadelphia firefighter and not long ago one opened up in Lincoln, NE.

Thanks to Richmond In Fire Chief Jerry Purcell, who I met during a visit to Richmond to do a Firehouse story on their 1968 explosion in downtown. As a result of

William, "Bill" Doty retired Hazmat Chief Philadelphia Fire Department.

the Richmond story being published I was able to locate and become friends with blast survivor Jack Bales. More recently I visited to do another story on their hazmat team and propane training. Thanks to new friend Ron Huffman who traveled to Richmond to conduct the propane training utilizing water injection to control liquid propane leaks. The article appeared in the September 2019 *Firehouse Magazine.*

Thanks to Tod Allen, Fire Chief in Crete Nebraska who I met when I was researching a train derailment in Crete for another friend Kent Anderson. We have become good friends. Tod is the apparatus operator on Truck one at Station 1 for the Lincoln Nebraska Fire Department. He invited me to come and ride with him, and many adventures later I still go there on a regular basis. I thank all of my friends past and present on "B" Shift at Station 1 for making me feel at home and showing me a good time whenever I am there. Thanks to friend Captain Mark Majors for sharing his experiences with Nebraska Task Force 1 Urban Search and Rescue Team (USAR) and Captain Francisco Martinez Lincoln Hazmat. Finally, I thank Chief Michael Despain and assistant Chief Patrick Borer for their friendship and hospitality while visiting the Lincoln Fire Department on many occasions. This is only the short list—I would have to write a separate book to thank all of you I have met and for the impact you have had on my life over the past 40+ years. You know who you are; I appreciate your friendship and assistance and consider your selves thanked again.

Chief Jerry Purcell Richmond, IN Fire Department.

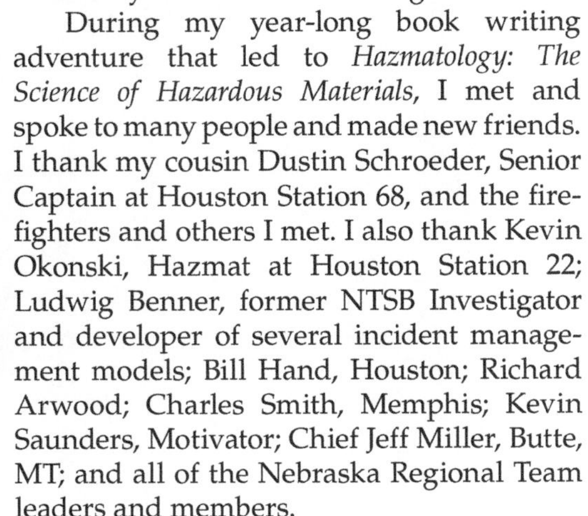

During my year-long book writing adventure that led to *Hazmatology: The Science of Hazardous Materials,* I met and spoke to many people and made new friends. I thank my cousin Dustin Schroeder, Senior Captain at Houston Station 68, and the firefighters and others I met. I also thank Kevin Okonski, Hazmat at Houston Station 22; Ludwig Benner, former NTSB Investigator and developer of several incident management models; Bill Hand, Houston; Richard Arwood; Charles Smith, Memphis; Kevin Saunders, Motivator; Chief Jeff Miller, Butte, MT; and all of the Nebraska Regional Team leaders and members.

I express my thanks to my cousin Jeanene and her husband Randy for coming all the way from Montana to be with me at the Firehouse Hall of Fame induction. I am also grateful to Brent Boydston, James Rey Milwaukee, Wilbur Hueser and Saskatoon in Canada for the hospitality and tour, and Captain Oscar Robles, Imperial, CA. The list just goes on and on, and there is no room here for everyone, but the rest of you know who you are and I want you to know how much your assistance is appreciated. You are all considered friends, and I hope we will talk and or meet again. Finally, thanks to librarians and historians across the country for your assistance in research, thanks for the memories!

Robert Burke

Author

Robert A. Burke was born in Beatrice and grew up in Lincoln, Nebraska; graduated from high school in Dundee, Illinois; and earned an AA in Fire Protection Technology from Catonsville Community College, Baltimore County, Maryland (now Community College of Baltimore County) and a BS in Fire Administration from the University of Maryland. He has also pursued his graduate work at the University of Baltimore in Public Administration. Mr Burke has attended numerous classes at the National Fire Academy in Emmitsburg, Maryland, and additional classes on firefighting, hazardous materials and Weapons of Mass Destruction at Oklahoma State University; Maryland Fire and Rescue Institute; Texas A & M University, College Station, Texas; the Center for Domestic Preparedness in Anniston, Alabama; and others.

Mr. Burke has over 40 years' experience in the emergency services as a career and volunteer firefighter, and has served as a Lieutenant for the Anne Arundel County, Maryland Fire Department; an assistant fire chief for the Verdigris Fire Protection District in Claremore, Oklahoma; Deputy State Fire Marshal in the State of Nebraska; a private fire protection and hazardous materials consultant; and an exercise and training officer for the Chemical Stockpile Emergency Preparedness Program (CSEPP) for the Maryland Emergency Management Agency; and retired as the Fire Marshal for the University of Maryland. He has served on several volunteer fire companies, including West Dundee, Illinois; Carpentersville, Illinois; Sierra Volunteer Fire Department, Chaves County, New Mexico; Ord, Nebraska; and Earleigh Heights Volunteer Fire Company in Severna Park, Maryland, which is a part of the Anne Arundel County, Fire Department, Maryland.

Mr. Burke has been a Certified Hazardous Materials Specialist (CFPS) by the National Fire Protection Association (NFPA) and certified by the National Board on Fire Service Professional Qualifications as a Fire Instructor III, Fire Inspector, Hazardous Materials Incident Commander, Fire Inspector III and Plans Examiner II. He served on the NFPA technical committee for NFPA 45 Fire Protection for Laboratories Using Chemicals for 10 years. He has been qualified as an expert witness for arson trials as well.

Mr. Burke retired as an adjunct instructor at the National Fire Academy in Emmitsburg, Maryland in April 2018 after 30 years. He taught Hazardous Materials, Weapons of Mass Destruction and Fire Protection curriculums. He taught at his Alma Mater Community College of Baltimore County, Catonsville Campus and Howard County Community College in Maryland. He has had articles published in various fire service trade magazines for the past 33 years. Mr. Burke is currently a contributing editor for *Firehouse Magazine*, with a bimonthly column titled "Hazmatology," and he has had numerous articles published in *Firehouse, Fire Chief, Fire Engineering* and *Nebraska Smoke Eater* magazines. He was inducted into the Firehouse Hall of Fame in October 2018 in Nashville, TN. Mr Burke has also been recognized as a subject matter specialist for hazardous materials and been interviewed by newspapers, radio and television about incidents that have occurred in local communities including Fox Television in New York City live during a tank farm fire on Staten Island.

Mr. Burke has been a presenter at Firehouse Expo in Baltimore, MD and Nashville, TN numerous times, most recently in 2017. He gave a presentation at the EPA Region III SERC/LEPC Conference in Norfolk, Virginia, in November 1994 and a presentation at the 1996 Environmental and Industrial Fire Safety Seminar, Baltimore, Maryland, on DOT ERG. He was a speaker at the 1996 International Hazardous Materials Spills Conference on June 26, 1996, in New Orleans, Louisiana; a speaker at the Fifth Annual1996 Environmental and Industrial Fire Safety Seminar in Baltimore, Maryland, sponsored by Baltimore City Fire Department; and at LEPC, an instructor for Hazmat Chemistry, August 1999, at Hazmat Expo 2000 in Las Vegas, Nevada. He also delivered a Keynote presentation at the Western Canadian Hazardous Materials Symposium Saskatoon, Saskatchewan, Canada, in 2008.

Mr. Burke has developed several CD-ROM-based training programs, including the Emergency Response Guide Book, Hazardous Materials and Terrorism Awareness for Dispatchers and 911 Operators, Hazardous Materials and Terrorism Awareness for Law Enforcement, Chemistry of Hazardous Materials Course, Chemistry of Hazardous Materials Refresher, Understanding Ethanol, Understanding Liquefied Petroleum Gases, Understanding Cryogenic Liquids, Understanding Chlorine and Understanding Anhydrous Ammonia. He has also developed the "Burke Placard Hazard Chart." He has published seven additional books titled *Hazardous Materials Chemistry for Emergency Responders 1st, 2nd and 3rd*

Editions, *Counterterrorism for Emergency Responders 1st, 2nd and 3rd editions*, *Fire Protection: Systems and Response* and *Hazmat Teams Across America*.

Currently, Mr. Burke serves on the Homestead LEPC in Southeast Nebraska. He also manages a Hazardous Materials section at the Nebraska Firefighters Museum and periodically rides with friends on "B" shift at Station 1, Lincoln Fire Department. He can be reached via email at robert.burke@windstream.net, on Facebook at https://www.facebook.com/RobertAb8731 and through his website: www.hazardousmaterialspage.com.

Volume Four

Common Sense Emergency Response

> Firemen do not consider themselves heroes because they do what the job requires.
>
> **Edward Croker Chief FDNY1908, speaking on the death of five firemen**

Hazardous materials incident scenes, like many emergency scenes, are often chaotic when emergency responders arrive (Figure 4.1). In a nutshell, our job is to bring order utilizing incident command, protect, and stabilize the situation. One of the most important things to remember when responding to hazardous materials incidents is that we did not cause the incident, but we do not want to make the situation worse than it would have been if we had not responded at all. Unfortunately, as history has

Figure 4.1 If we do not recognize when hazardous materials are present, nothing else matters. (Courtesy: Paul Combs.)

shown us, there have been times where responders inadvertently did make things worse. Not because that was the intention but rather by doing something that should not have been done. None of the actions were done intentionally, but because responders did not know any better at the time. Being an emergency responder in the 21st century gives us the ability to look back and see where we were when Ron Gore and his team members in Jacksonville, FL were blazing the trails of hazmat response. If you look at their accomplishments without much to go on and yet they had no serious injuries or deaths.

Today we have laws, regulations, standards, and information literally at our finger tips. We have the benefits of lessons learned from those who came before us. Yet, we still lose the lives of emergency responders dealing with hazardous materials in the same manner as we did decades ago. As recently as 2018 a firefighter lost his life in a fire involving a propane explosion. Within the last decade, multiple emergency responders were killed fighting fires involving ammonium nitrate fertilizer. Members of the public have lost their lives in anhydrous ammonia incidents. Not one emergency responder has lost their life at an "exotic" hazardous materials incident. Statistically, it is common everyday hazardous materials that escape their containers on a regular basis. (Figure 4.2). Propane, anhydrous ammonia, ammonium nitrate, chlorine, ethanol, crude oil, acids, bases, and 50% of the time, hydrocarbon fuels.

Figure 4.2 Statistically, it is common everyday hazardous materials that escape their containers on a regular basis. (Courtesy: Livingston, LA Fire Department.)

We owe it to those who lost their lives pioneering hazardous materials response to ensure they did not lose them in vain if we can learn from the consequences of their misguided actions.

It is time to get back to the basics and spend our time preparing for what we know will happen rather than what everyone tells us could happen. If we have the basic skills to deal with what does happen and handle those incidents well, we can deal with whatever else might happen. Plan, train, and practice for materials that are in our jurisdictions. In my view, there is a vast difference between hazardous materials exposure from one part of the country to another. Yet, we try to prepare for everything including terrorism, everywhere.

Rail accidents are rare, over 99% of all hazardous materials shipped by rail get to their destination safely. Highway traffic, on loading, or off loading accidents involving tankers are more frequent, but once again the catastrophic highway incidents are pretty rare. Industrial incidents occur more frequently than any other type; however, career events do not happen very often. The U.S. Chemical Safety Board is charged with investigating fixed facility accidents involving hazardous materials. Since their inception 20 years ago, they have only investigated 133 major incidents (Figure 4.3). That averages about seven each year.

Having spent time traveling around the country visiting and writing about hazmat teams and incidents for the past 33 years, I have found that some things are different and some things are the same. Regardless of the hazmat exposures, all responders need to have the same basic skills. Not all responders, however, need the same tools in their "Hazmat Tool Box". Why do responders need to know about and have the tools to deal with railroad cars if they do not live in an area where there are trains? The same goes for water transportation, pipeline transportation, and fixed storage and use. What we do need is the knowledge of and the tools to deal with those hazardous materials we know will escape their containers in our response area.

Another issue that I believe is critical is the way we respond to hazardous materials incidents. Hazardous materials incidents and fire response are not the same. Throughout the history of the fire service we have learned to get to a scene of a fire quickly, but safely. When we arrive we prepare and attack the fire as fast as possible to save lives and property. Unfortunately, we sometimes do so without regard for our own safety and well being. Firefighting is without a doubt a dangerous profession. Firefighters run toward an emergency as others are running to escape. If we approach firefighting in a more scientific way, we can reduce the dangers and take realistic actions safely. I will stop there, because this is not a book about firefighting, but rather hazardous materials response.

Figure 4.3 Since their inception 20 years ago, they have investigated 133 major incidents. (From CSB.)

The point I am trying to make is when responding to hazardous materials incidents we need to SLOW DOWN. Take the time to scientifically determine what the hazards are, what the threats to the public and responders are, and what is to be gained by whatever actions we decide upon. Use some common sense. What are the risks and what are the benefits of our taking action or taking no action. Yes, I know that may be a hard pill to swallow, but when hazardous materials escape, the best action may be no action at all other than getting the public and responders out of harm's way. Risking responders lives to save property and the environment should not be an option.

Developing an Emergency Response Plan

Developing emergency response plans (ERP) that focus on the chemicals stored, used, and transported within our communities will benefit everyone. Emergency response agencies are required under the Emergency Planning and Community Right-to-Know Act (EPCRA), Occupational Safety and Health Administration (OSHA), and Environmental Protection Agency (EPA) to have plans for response to hazardous materials incidents. Besides these requirements, it is just common sense to plan for response to incidents that require a large amount of technical knowledge for effective and safe handling. Planning is a team effort. Every agency or group of agencies that have a stake should be involved in the planning process.

> ***Author's Note:*** *I was talking to an Emergency Manager about their emergency plan. He was so proud of his plan and glad having completed it. However, he complained to me that he couldn't get the response agencies in the community to buy into the plan. My first question to him "was everyone in the response community involved in the plan development?" His response was "no, I just wrote the plan myself." That is why no one would buy into the plan. Everyone with a stake needs to be involved in the planning process.*

OSHA CFR 1910.120 (q)(2)

Elements of an emergency response plan. The employer shall develop an emergency response plan for emergencies which shall address, as a minimum, the following areas to the extent that they are not addressed in any specific program required in this paragraph:

- Pre-emergency planning and coordination with outside parties.
- Personnel roles, lines of authority, training, and communication.
- Emergency recognition and prevention.
- Safe distances and places of refuge.
- Site security and control.
- Evacuation routes and procedures.
- Decontamination.
- Emergency medical treatment and first aid.
- Emergency alerting and response procedures.
- Critique of response and follow-up.
- PPE and emergency equipment.

Emergency response organizations may use the local emergency response plan or the state emergency response plan or both, as part of their emergency response plan to avoid duplication. Those items of the emergency response plan that are being properly addressed by the SARA Title III plans may be substituted into their emergency plan or otherwise kept together for the employer and employee's use (OSHA).

Step 1: Form a Collaborative Planning Team

Hazardous materials planning should come out of a process coordinated by a team. The team is the best vehicle for incorporating the expertise of various sources into the planning process and for producing an accurate and complete document. In selecting the members of a team who will have overall responsibility for hazardous materials planning, the following four considerations are the most important:

- The members of the group must have the ability, commitment, authority, and resources to get the job done;
- the group must possess, or have ready access to, a wide range of expertise related to the community, its industrial facilities and transportation systems, and the mechanics of the emergency response and response planning;
- the members of the group must agree on their purpose and be able to work cooperatively with one another; and
- the group must be representative of all elements of the community with a substantial interest in reducing the risks posed by hazardous materials.

In organizing the community to address the problems associated with hazardous materials, it is important to bear in mind that all affected parties have a legitimate interest in choosing from the planning alternatives. Therefore, strong efforts should be made to ensure that all groups with an interest in the planning process are included. For the planning team to function effectively, its size should be limited to a workable number.

In communities with many interested parties, it will be necessary to carefully select from them to ensure fair and comprehensive representation. Some individuals may feel left out of the planning process. People should be given access to the process through the various approaches discussed in the following sections such as membership on a task force or advisory council. In addition, all interested parties should have an opportunity to provide their input during the review process.

Step 2: Understand the Situation

When the planning team members and their leaders have been identified and a process for managing the planning tasks is in place, the team should address several interrelated tasks.

- Review of existing plans, which prevents plan overlap and inconsistency, provides useful information and ideas, and facilitates the coordination of the plan with other plans.
- Review of the Risk Management Plan (RMP) information, which includes reviews of RMPs, offsite consequence analysis data, and local facility plans.
- Hazards analysis, which includes identification of hazards, analysis of vulnerability, and analysis of risk.
- Assessment of preparedness, prevention, and response capabilities, which identifies existing prevention measures and response capabilities (including mutual aid agreements), and assesses their adequacy.
- Completion of hazardous materials planning that describes the personnel, equipment, and procedures to be used in case of accidental release of a hazardous material.
- Development of an ongoing program for plan implementation, maintenance, training, and exercising.

Step 3: Determine Goals and Objectives

- **Goals** are general guidelines that explain what you want to achieve in your community. They are usually long-term and represent global visions such as "protect public health and safety."
- **Objectives** define strategies or implementation steps to attain the identified goals. Unlike goals, objectives are specific, measurable, and have a defined completion date. They are more specific and outline the "who, what, when, where, and how" of reaching the goals.

Step 4: Plan Development

A. Introduction
  1. Incident Information Summary
  2. Promulgation Document
  3. Legal Authority and Responsibility for Responding
  4. Table of Contents
  5. Abbreviations and Definitions
  6. Assumptions/Planning Factors
  7. Concept of Operations
     a. Governing Principles
     b. Organizational Roles and Responsibilities
     c. Relationship to Other Plans (community-wide or installation specific)

    8. Instructions on Plan Use
        a. Purpose
        b. Plan Distribution
    9. Record of Amendments
B. Emergency Assistance Telephone Roster
C. Response Functions
    1. Initial Notification of Response Agencies
    2. Direction and Control
    3. Communications (among Responders)
    4. Warning Systems and Emergency Public Notification
    5. Public Information/Community Relations
    6. Resource Management
    7. Health and Medical Services
    8. Response Personnel Safety
    9. Personal Protection of Citizens
        a. Indoor Protection (Shelter in Place)
        b. Evacuation Procedures
        c. Other Public Protection Strategies (e.g., Plume Suppression, Containment)
    10. Fire and Rescue
    11. Law Enforcement
    12. Ongoing Incident Assessment
    13. Human Services
    14. Public Works
    15. Others
D. Containment and Clean up
    1. Techniques for Spill Containment and Clean up
    2. Resources for Clean up and Disposal
E. Documentation and Investigative Follow-up
F. Procedures for Testing and Updating Plan
    1. Testing the Plan
    2. Updating the Plan
G. Hazards Analysis (Summary)
H. References
    1. Laboratory, Consultant, and Other Technical Support Resources
    2. Technical Library

Step 5: Preparation, Review, and Approval

Write the Plan

This step turns the results of game planning into an emergency plan. The planning team develops a rough draft of the base plan, functional or hazard or annexes, or other parts of the plan as appropriate. The recorded results of the game planning process used in the previous step provide an outline for the rough draft. As the planning team works through successive drafts, they add necessary tables, charts, and other graphics. A final draft is prepared and circulated to organizations that have the responsibility of implementing the plan for their comments.

Following these simple rules for writing plans and procedures will help ensure that readers and users understand the content.

- Keep the language simple and clear by writing in plain English.
- Summarize important information using checklists and visual aids such as maps and flowcharts.
- Avoid using jargon.
- Use short sentences and active voice. Qualifiers and vague words only add to confusion.
- Provide enough detail to convey an easily understood concept of operations. The less certain a situation, the less detail can be put into the plan. Those parts of a plan that would be most affected by the hazard should have the least amount of detail. Conversely, those that would be least affected by the hazard should have the most amount of detail. The amount of detail a plan should provide depends on the target audience and the amount of certainty about the situation. Similarly, plans written for a jurisdiction or organization with high staff turnover might require more detail.
- Format the plan and present its content so that the readers can quickly find solutions and options. Focus on providing mission guidance and not on discussing policy and regulations. Plans should provide guidance for carrying out common tasks as well as provide enough insight into intent and vision so that responders can handle unexpected events. However, when writing a plan, "stay out of the weeds." Procedural documents (e.g., standard operating procedures) should provide the fine details.

Approve the Plan

The written plan should be checked for its conformity to applicable regulatory requirements and the standards of Federal or State agencies (as appropriate), as well as for its usefulness in practice. Planners should consult the next level of government about its emergency plan review cycle. Reviews of plans allow other agencies with emergency responsibilities to suggest improvements based on their accumulated experience.

Step 6: Plan Implementation and Maintenance

The ERP should:

- outline the triggers and procedures for plan implementation;
- specify communication pathways and identify emergency resources;
- provide strategies for evaluating plan effectiveness in response to an incident and making improvements;
- delineate the plan review, approval, and update process; and
- define the ERP training program.

The ERP is a dynamic document that must be reviewed periodically and revised to incorporate changes to critical components such as essential personnel or outside contact information, or updates from vulnerability analyses or after-incident reviews. A revision log should be kept in front of the ERP document. The latest approved version should be distributed, with initial or refresher training conducted to ensure all essential response personnel are familiar with the ERP, their roles/responsibilities, and resources available during an emergency.

OSHA identifies additional specific planning requirements for hazmat emergency response within a given jurisdiction or multiple jurisdictions (OPM.gov):

Training

Once the emergency response plan is in place, emergency response personnel should be trained to the appropriate levels according to what their functions will be on the scene of a hazardous materials incident (Figure 4.4).

> ***Author's Note:*** *OSHAs basic premise on training is that you train and equip your personnel for the job you are asking them to do. With that in mind, the areas between operations and technicians can become fuzzy. For example, operations-level personnel can be trained to do decontamination freeing up technicians to focus on hot zone issues.*

Trainers

Trainers who teach any of the above training subjects shall have satisfactorily completed a training course for teaching the subjects they are expected to teach, such as the courses offered by the U.S. National Fire Academy, or they shall have the training and/or academic credentials and instructional experience necessary to demonstrate competent instructional skills, as well as a good command over the subject matter of the courses they are to teach.

Figure 4.4 Once the emergency response plan is in place, emergency response personnel should be trained to appropriate levels according to what their functions will be on the scene of a hazardous materials incident. (Courtesy: Gwinnett County Georgia Police Department.)

OSHA Training Levels CFR 1910.120(q)

OSHA regulations dictate the levels of training for emergency responders to hazardous materials releases in CFR 1910.120(q). Training shall be based on the duties and functions to be performed by each responder of an emergency response organization. The skill and knowledge required for all new responders, those hired after the effective date of this standard, shall be conveyed to them through training before they are permitted to participate in actual emergency operations on an incident. Employees who participate, or are expected to participate, in emergency response shall be given training in accordance with the following.

Awareness

First responders at the awareness level are individuals who are likely to witness or discover a hazardous substance release and who have been trained to initiate an emergency response sequence by notifying the proper authorities of the release. These include personnel from law enforcement, public works, department of roads, industry, and others. Awareness does not apply to fire, emergency medical services, and hazmat response (Figure 4.5). Awareness-level personnel would take no further action beyond notifying the authorities of the release. First responders at the awareness level shall have sufficient training or have had sufficient experience to objectively demonstrate competency in the following areas:

Figure 4.5 Awareness does not apply to fire, emergency medical services, and hazmat response.

- Understanding of what hazardous substances are and the risks associated with them during an incident.
- Understanding of the potential outcomes associated with an emergency created when hazardous substances are present.
- Ability to recognize the presence of hazardous substances in an emergency.
- Ability to identify hazardous substances, if possible.
- Understanding of the role of the first responder awareness personnel in the employer's emergency response plan, including site security and control and the U.S. Department of Transportation's Emergency Response Guidebook.
- Ability to realize the need for additional resources and to make appropriate notifications to the communication center.

Operations

First responders at the operations level are individuals who respond to releases or potential releases of hazardous substances as part of the initial response to the site to protect nearby persons, property, or the environment from the effects of the release. They are trained to respond in a defensive manner without actually trying to stop the release. Their function is to contain the release from a safe distance, keep it from spreading, and prevent exposure. First responders at the operational level (Figure 4.6) shall have received at least 8 h of training or have had sufficient experience to objectively demonstrate competency in the following areas in addition to those listed for the awareness level, which the employer shall certify.

Figure 4.6 First responders at the operational level are generally limited to there firefighting PPE as they arrive on scene.

- Knowledge of the basic hazard and risk assessment techniques.
- Knowledge of how to select and use proper personal protective equipment provided to the first responder at the operational level.
- Understanding of basic hazardous materials terms.
- Knowledge of how to perform basic control, containment, and/or confinement operations within the capabilities of the resources and personal protective equipment available with the unit.
- Knowledge of how to implement basic decontamination procedures.
- Understanding of the relevant standard operating procedures and termination procedures.

OSHA Formal Interpretation of Awareness/Operations Level

According to OSHA, awareness-level personnel are those who, in the course of their normal duties, could encounter an emergency involving hazardous materials and who are expected to recognize the presence of hazardous materials, protect themselves, call for trained personnel, and secure the area. An example would be department of roads or public works personnel.

When emergency response organizations respond to a hazmat incident, OSHA stated in a formal interpretation (July 25, 2007) that those responders should be operations-level personnel who first arrive at the scene of the incident. Operations-level personnel are those who respond to a hazmat incident to protect nearby persons, environment, or property from the effects of the release. Operations-level personnel are required to have awareness training as well as an additional 8h of operations-level training. Awareness- and operations-level personnel do not enter the "hot" zone.

Technician

Hazardous materials technicians are individuals who respond to releases or potential releases for the purpose of stopping the release (Figure 4.7). They assume a more aggressive role than a first responder at the operations level in that they approach the point of release to plug, patch, or otherwise stop the release of a hazardous substance. Hazardous materials technicians shall have received at least 24hours of training equivalent to the first responder operations level and additionally have competency in the following areas which the employer shall certify.

> ***Hazmatology Point***: *OSHA's requirement of 24 hours is not realistic in terms of the amount of training necessary for hazardous materials technicians to perform the vast number of duties required at a hazardous materials incidents. This is the minimum OSHA requires your personnel*

Figure 4.7 Hazardous materials technicians are individuals who respond to releases or potential releases for the purpose of stopping the release. (Courtesy: Sparks, NV Fire Department.)

to have. Your department will need to determine what amount of training is required and what subject areas to ensure that your personnel can operate safely and effectively at a hazardous materials incident.

- Knowledge of how to implement the employer's emergency response plan.
- Knowledge of the classification, identification, and verification of known and unknown materials using field survey instruments and equipment.
- Ability to function within an assigned role in the incident command system.
- Knowledge of how to select and use proper specialized chemical personal protective equipment provided to the hazardous materials technician.
- Understanding of hazard and risk assessment techniques.
- Be able to perform advance control, containment, and/or confinement operations within the capabilities of the resources and personal protective equipment available with the unit.
- Ability to understand and implement decontamination procedures.
- Understanding of termination procedures.
- Understanding of basic chemical and toxicological terminology and behavior.

Specialist

Hazardous materials specialists are individuals who respond with and provide support to hazardous materials technicians. Their duties parallel those of the hazardous materials technicians, however, they require a more direct or specific knowledge of the various substances they may be called upon to contain. The hazardous materials specialist would also act as the site liaison with federal, state, local, and other government authorities in regards to site activities. Hazardous materials specialists shall have received at least 24h of training equivalent to the technician level, and additionally have competency in the following areas, which the employer shall certify.

- Knowledge of how to implement the local emergency response plan.
- Understanding of the classification, identification, and verification of known and unknown materials using advanced survey instruments and equipment.
- Knowledge of the state emergency response plan.
- Ability to select and use proper specialized chemical personal protective equipment provided to the hazardous materials specialist.
- Understanding of in-depth hazard and risk techniques.

- Ability to perform specialized control, containment, and/or confinement operations within the capabilities of the resources and personal protective equipment available.
- Ability to determine and implement decontamination procedures.
- Ability to develop a site safety and control plan.
- Understanding of chemical, radiological, and toxicological terminology and behavior.

Incident Commander

On-scene incident commanders (Figure 4.8), who will assume control of the incident scene beyond the first responder awareness level, shall receive at least 24 h of training equivalent to the first responder operations level and additionally have competency in the following areas, which the employer shall certify.

- Knowledge of and ability to implement the employer's incident command system.
- Knowledge of how to implement the employer's emergency response plan.
- Knowledge and understanding of the hazards and risks associated with employees working in chemical protective clothing.
- Knowledge of how to implement the local emergency response plan.
- Knowledge of the state emergency response plan and the Federal Regional Response Team.
- Knowledge and understanding of the importance of decontamination procedures.

Figure 4.8 On-scene incident commander. Incident commanders. (Courtesy: Saint Louis County Fire Department.)

Refresher Training

Employees who are trained in accordance with paragraph (q)(6) of 29 CFR 1910.120 shall receive annual refresher training of sufficient content and duration to maintain their competencies, or shall demonstrate competency in those areas at least yearly (OSHA).

First Communication of Hazardous Materials Incident

Dispatchers and 911 operators are usually the first to receive notification of any type of emergency in the community (Figure 4.9). They may be talking with the person who is a victim of an incident, who witnessed what happened, or who may even be the perpetrator. Callers can provide valuable information to assist emergency responders. Dispatchers and 911 operators should be trained to provide them the tools to extract appropriate information from callers. Nearly every fire and EMS response has the potential for the presence of hazardous materials. Call takers should use their knowledge and training to gather, based upon the caller's information, more specific information about hazardous materials which may be present.

It Begins with a Call

Dispatchers and 911 operators or any other type of call taker need to be able to determine when a call may involve hazardous materials. They play one of the most important roles in the emergency response process. Call

Figure 4.9 Dispatchers and 911 operators are usually the first to receive notification of any type of emergency, which occurs in the community.

takers are the first to talk with someone who has information about an emergency. If they are making the call, they likely witnessed what happened or at least what prompted them to call 911 in the first place. Training call takers to be able to recognize the presence of hazardous materials is at least equal to that of training first responders in the field.

Preparing Dispatchers for Hazmat & Terrorism

Gathering Information

Call takers should routinely ask questions about the presence of hazardous materials when emergency calls come in. This includes traffic accidents, industrial accidents, EMS, and fire responses. Training should further prepare call takers to provide assistance to emergency responders when dealing with hazardous materials, terrorist, or threat incidents. Threat incidents may include bomb threats, anthrax threats, or any other type of threat involving explosive, chemical, or biological materials. Through training, dispatchers will develop a better understanding of the type of information that is important to response personnel. They will also understand the terminology used by response personnel and will be able to provide a higher level of assistance based upon their increased level of knowledge. Hazardous materials, terrorist, and threat incidents are very technical in nature. Training will help increase the awareness level of dispatchers and 911 operators when these incidents occur.

Information provided is applicable to fire, police, and EMS dispatchers. Topics that should be covered in a dispatcher and 911 operator training course should include an introduction to hazardous materials incidents and acts of terrorism. Information should include hazardous materials and terrorist awareness; hints to the presence of hazardous materials or terrorist incidents; locations where such incidents may occur; recognizing signs and symptoms from victims; and the type of information to gather from the caller that may be helpful to response personnel.

Every community should have a plan for dealing with hazardous materials releases or terrorist incidents. Call takers should be thoroughly familiar with the plan and know when and how to implement the plan if needed. Standard Operating Procedures (SOPs) based upon the community emergency plan and dispatch operations should be developed for day-to-day use. SOPs should cover call screening, initial dispatch, when the plan is activated, who is in charge, hazmat references, dealing with the media, and dispatcher duties. SOPs should also clearly define the role of the dispatcher during a hazardous materials or terrorist incident.

Dispatchers and 911 operators should keep the caller on the line as long as possible to obtain information to pass on to emergency responders. Initial dispatch protocols should include proper resources, weather conditions, information about the hazard or threat, and concise information about what may have happened and what the conditions are at the present time. Call takers will have numerous duties and responsibilities during a hazardous materials or terrorist incident. They do not end with the dispatch of emergency responders. Dispatchers should have access to reference materials including the Emergency Response Guide Book.

Other important references include electronic information sources as well as hard copy like the Condensed Chemical Dictionary, the NIOSH Guide, Bureau of Explosives Railroad Book, among others. Dispatchers, who obtain proper information or fail to recognize when the potential for hazardous materials or terrorism is present, can mean the difference between a successful outcome and a disaster. Fire departments respond to a large numbers of emergency medical calls on a daily basis. Dispatchers and 911 operators should be alert for unusual events. Multiple medical calls from different locations and victim's exhibiting similar symptoms may indicate an unusual problem.

Mass casualty incidents where no trauma has occurred are also unusual and should be recognized by the dispatcher as an unusual event. Reports of an explosion with little damage is also a "red flag." Call takers should be familiar with information that would indicate a potential terrorist incident. Members of the media listen to dispatch frequencies to get a heads-up on emergency incidents in the community. Dispatchers should be prepared to field calls from the media about what has happened, what is the local response, is anyone killed or injured, is there an evacuation, or is anyone in the community threatened?

Location of an event can be helpful information in terms of potential exposure and response difficulties. Is the emergency a transportation event or fixed facility? Extract as much information as possible from the caller about the location and surrounding exposure, topography, rural versus urban setting, population exposure, etc. Time of day or day of the week can be important in terms of weather stability, effect on vapor clouds, traffic flow, and population densities. Did the incident just happen, or was there a delay in reporting? Are you receiving multiple calls about the incident or multiple calls with different locations?

What event occurred, was it a fire, explosion, vehicle accident, medical emergency, natural disaster, criminal act, act of terrorism, or chemical release? Fires can result in explosions, and explosions can cause fires. If an explosion has occurred, there could be a secondary device present. The dispatcher should alert responders to potential secondary devices. Natural disasters as well as criminal and terrorist acts can result in the release of hazardous materials. Once again, the information the call taker extracts from the person reporting the incident is critical in determining what has occurred. Incidents involving hazardous materials can be accidental or criminal acts.

Terrorist events are almost always criminal acts. How the incident happened can provide responders with information about what they may face when they arrive. Was the incident an accident? Was the incident intentional? Incidents can also happen as a result of negligence. Developing plans, SOPs, and providing training for dispatchers is key if an incident outcome is to be successful. The role of the dispatcher is critical and should not be overlooked or underestimated.

Identifying incident levels can assist responders and dispatchers in determining what resources will be required to handle an incident. Level I incidents are small scale and may be handled by the first responders with one company. It is unlikely that evacuation will be necessary, and injuries are unlikely unless it is a vehicle accident. Level II incidents are larger, requiring multiple companies, hazardous materials team, and, depending on the department size, mutual aid. Evacuations or sheltering in place is likely to occur. There may be injuries or even deaths. Level III incidents are catastrophic community emergencies. They are well beyond the local ability to deal with them. Large evacuations will be likely and maybe for an extended period of time. State and Federal resources may be required. Injuries and deaths are likely. Developing resource contacts for various incident levels can streamline the notification process.

Once an incident has been identified as a hazardous materials or terrorist incident, responders should request the resources needed to bring the emergency safely under control. Dispatchers will be making the notifications and should have SOPs and copies of local emergency plans to contact the appropriate resources. The agencies that need to be notified should be outlined in the plan. These include local, state, and Federal agencies and private industry. Local emergency plans and SOPs should determine who gets called based upon the level of incident identified by the dispatcher from the initial call or the first arriving emergency responders. The plan should include organization or agency name, emergency contact phone number, and contact information for state and federal resources.

Information needed to make notifications for a hazardous materials incident can be placed into three general categories: organization/agency, emergency response telephone number, and resources available. Local and state notifications vary depending on the jurisdiction and its resources. Local and state departments of environment as well as health departments may need to be called. National notifications include CHEMTREC, a service of the Chemical Manufacturers Association, which can be contacted 24 hours a day at 1-800-424-9300. CHEMTREC is the contact point for industry, the shipper, and the manufacturer. If the material spilled is a marine pollutant or if oil products are spilled on water, the National Response Center (NRC) should be notified. NRC also operates 24 hours a day. (*Firehouse Magazine*).

Outside Resources

The NRC is operated by the U.S. Coast Guard, which is under the jurisdiction of the Department of Homeland Security (DHS) during peacetime. During time of war they are under the U.S. Navy. NRC is the notification, communications, technical assistance, and coordination center for the National Response Team (NRT). The NRC can provide chemical information much like CHEMTREC through their OM-TADS database. They are also the contact point to obtain federal assistance for hazardous materials incidents. If jurisdictions plan to seek reimbursement at any time under the U.S. Environmental Protection Agency's (EPA) reimbursement program, they must notify the NRC within 24 hours of the incident. The 24-hour emergency number for the NRC is 1-800-424-8802. NRC should also be contacted to report chemical or biological terrorist attacks. DOT Emergency Response Guide (ERG) provides information day or night on emergencies involving military shipments.

There are two contact numbers provided, for explosives or ammunition incidents call 703-697-0218, collect calls are accepted, all other dangerous goods incidents should be referred to 1-800-851-8061. These numbers are for emergencies only. The regional EPA response team, the Coast Guard, hazardous materials teams, and others identified locally should also be notified. Fixed facilities that have reportable quantities of extremely hazardous substances must have a facility emergency plan with 24-hour emergency contact numbers. This information can be obtained from the Local Emergency Planning Committee (LEPC). Other facilities may be contacted or preplanned to obtain after-hours emergency contact information.

Call takers should have a level of hazardous materials and terrorist incident knowledge that approaches the first responder awareness level. There are many clues that can assist the call taker in extracting the appropriate information from an emergency call. Much can be determined by the occupancy or use of a building or location. General knowledge of building use can help call takers determine if hazardous materials are likely to be present. Knowledge of the DOT placard and labeling system; NFPA 704 and other marking systems; transportation and fixed facility containers; the Emergency Response Guide Book and other references; shipping papers; Material Safety Data Sheets; and locations and indicators that a terrorist incident may have occurred are all critical for the emergency call taker.

Unlike hazardous materials incidents, there are few, if any, obvious clues as to the occurrence of a terrorist act. Call takers need to be aware of the types of buildings and locations that may be terrorist targets. Terrorists may also target high-profile events such as fairs and community celebrations or any event where large numbers of people may gather. Terrorists have also used attacks to commemorate a controversial event that has occurred previously. The bombing of the Federal Building in Oklahoma City occurred 2 years to the day of the Waco, Texas siege, which ended in the deaths of over 80 people. Dispatchers should be aware of on-scene indicators, which could indicate that a terrorist act has occurred. Victims may be the primary clue as to what has happened. Be alert for unusual symptoms and incidents where there are mass casualties without signs of trauma. Terrorists are looking for targets that are vulnerable.

Locations where little preparation or security is provided, "soft targets," are prime locations for terrorism. Terrorists look for public and government buildings, those with law enforcement or Internal Revenue Service (IRS) offices, infrastructure, such as power plants, tunnels, and bridges, water treatment plants, and hospitals, to name a few. Places with high economic impact, such as university research or medical facilities, financial institutions, the stock market, or commodities exchange could be potential targets. Telecommunications facilities, transmission towers, mass transit systems, and places with historical or symbolic significance might also be targeted. Government agencies provide critical services to the public, water supply, electricity in some cases, emergency services, environmental protection, health services, and many others. Much of our infrastructure is government-related, constructed, or maintained. There are many government buildings including schools, legislative, and administrative offices that terrorists may take issue with. Not all government agencies are popular; court systems, tax collection, probation, and others could be targeted.

Decision Trees

Hazardous materials and terrorist incidents do not happen on a regular basis. Call takers may become rusty with procedures because of lack of use. Decision trees can be useful in helping dispatchers determine what resources are needed and what questions to ask of the caller. Decision trees or flow charts, as they are sometimes called, are excellent tools for assisting dispatchers in determining what resources to allocate to an emergency call. Decision trees may be different from one community to another depending on the local resources. Decision trees start with a decision that needs to be made. In the case of emergency response, it is what resources need to be dispatched for certain types of hazardous materials and terrorist incidents.

Decision Tree Construction Steps

- Determine the decision to be made
- Ask questions that will lead to a decision
- Describe actions that will take place
- Determine when the process has ended

Decisions may also determine what type of incident has occurred. Decision trees can be constructed to assist the call taker in determining if hazardous material is present or if a terrorist incident has occurred. To come to an appropriate decision, you need to ask the proper questions. Take the time to ensure the questions asked would lead to the answer you are looking for. If not, you will need to revise the questions. Questions should eventually lead to actions. Based upon the information you collect from the caller and feed into the decision tree, you should be able to come up with an appropriate determination of the type of emergency and dispatch the appropriate resources. While this may all sound complicated, it is actually quite easy once you work through a few sample decision trees.

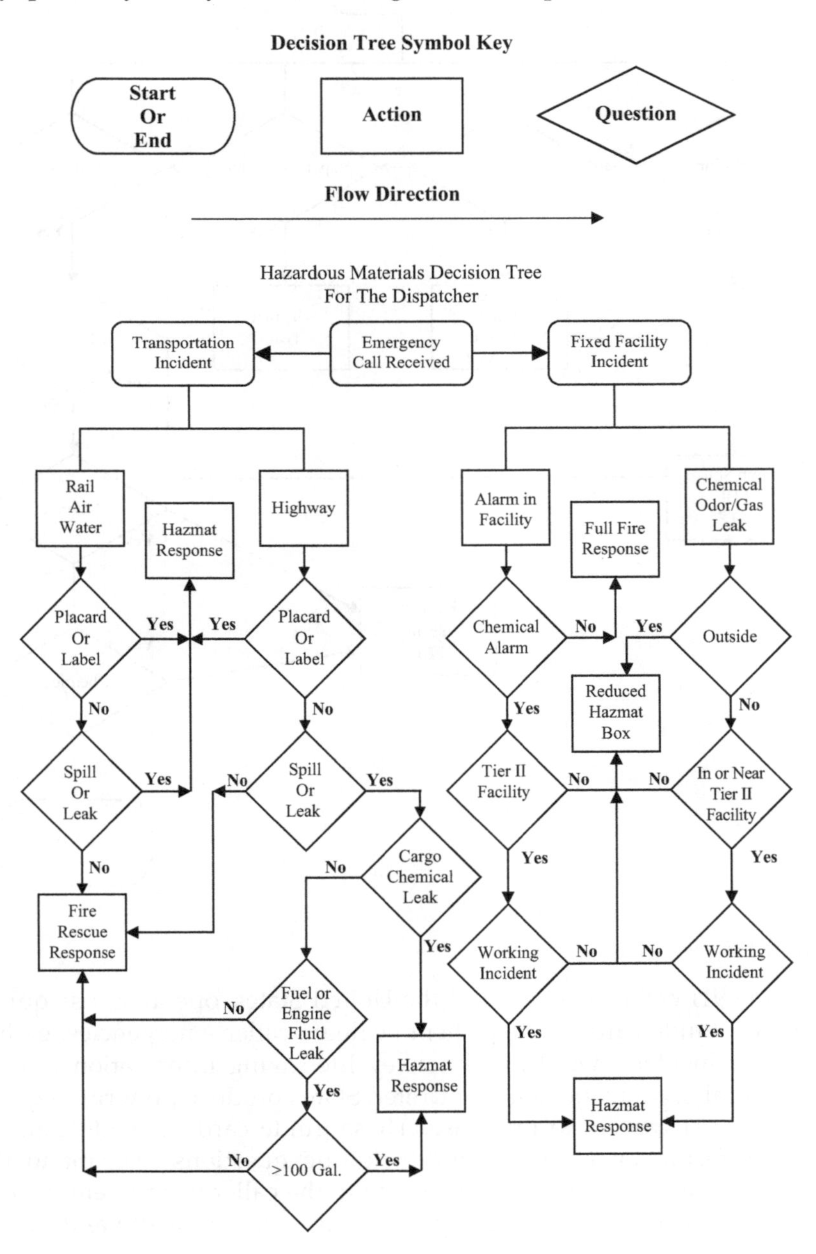

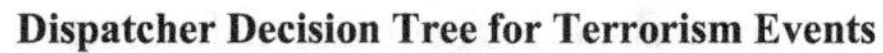

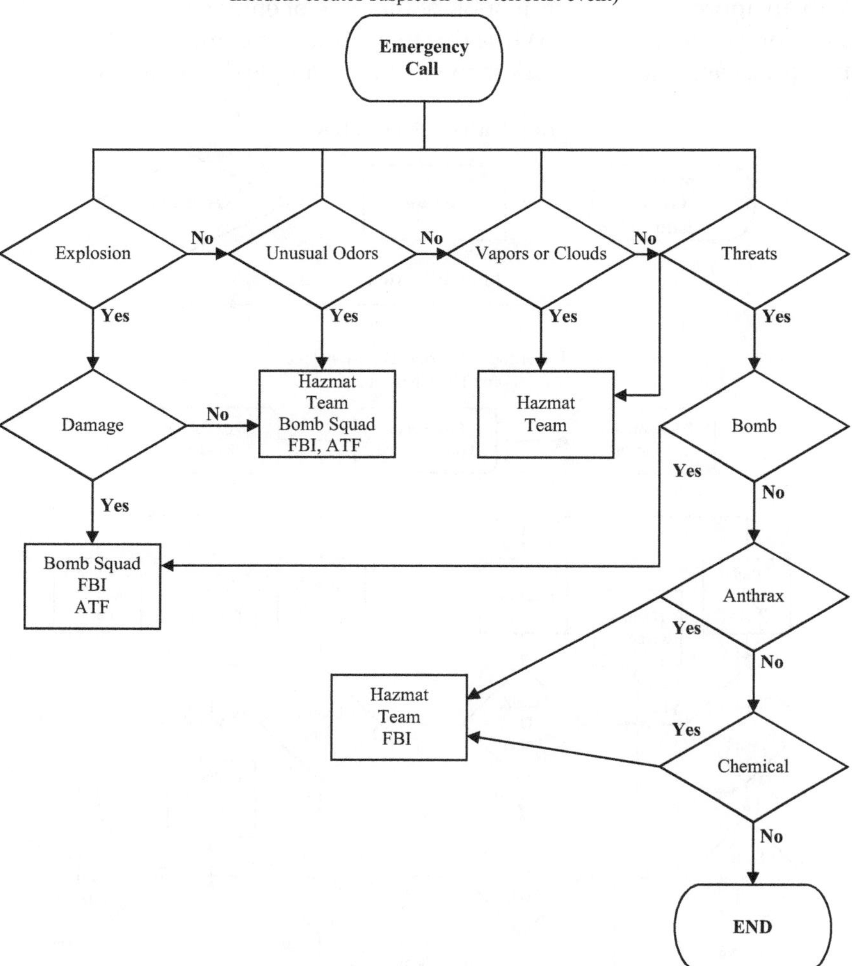

Checklists

At many 911 call centers around the United States, operators use quick-reference guide cards to help them evaluate caller emergencies, gather pertinent incident details, and convey life-saving information to callers. Several organizations in the United States produce prewritten guide cards for 911 centers to purchase. These guide cards provide easy-to-access information for operators, including questions relevant to the emergency and instructions to be given to the caller prior to emergency services' arrival. Specific sets of guide cards exist for health and injury,

Department of the treasury
Bureau of Alcohol, Tobacco & Firearms
BOMB THREAT CHECKLIST

1. When is the bomb going to explode?
2. Where is the bomb right now?
3. What does the bomb look like?
4. What kind of bomb is it?
5. What will cause the bomb to explode?
6. Did you place the bomb?
7. Why?
8. What is address?
9. What is your name?

EXACT WORDING OF BOMB THREAT:

Sex of caller: _____ Race: _____

Age: _____ Length of call: _____

Telephone number at which call is received: ________

Time call received: _____

Date call received: ___________

CALLER'S VOICE

☐ Calm	☐ Nasal
☐ Soft	☐ Angry
☐ Stutter	☐ Loud
☐ Excited	☐ Lisp
☐ Laughter	☐ Slow
☐ Rasp	☐ Crying
☐ Rapid	☐ Deep
☐ Normal	☐ Distinct

ATF F 1613.1 (Formerly ATF F 1730.1, which still may be used) (6-97)

☐ Slurred	☐ Whispered
☐ Ragged	☐ Clearing Throat
☐ Deep Breathing	☐ Cracking Voice
☐ Disguised	☐ Accent

☐ Familiar *(If voice is familiar, who did it sound like?)* ______________________

BACKGROUND SOUNDS:

☐ Street noises	☐ Factory machinery
☐ Voices	☐ Crockery
☐ Animal noises	☐ Clear
☐ PA System	☐ Static
☐ Music	☐ House noises
☐ Long distance	☐ Local
☐ Motor	☐ Office machinery
☐ Booth	☐ Other *(Please specify)*

BOMB THREAT LANGUAGE:

☐ Well spoken (education)	☐ Incoherent
☐ Foul	☐ Message read by threat maker
☐ Taped	☐ Irrational

REMARKS:

Your name:

Your position:

Your telephone number:

Date checklist completed: ___________

ATF F 1613.1 (Formerly ATF F 1730.1) (6-97)

fire service, and police response emergencies. Raleigh County 911 uses guide cards for health emergencies and injuries only. At 10:40 a. m. on the day of the incident, the propane service technician called Raleigh County 911 to report the release and summon emergency services. The operator who received the call did not have a guide card or protocol to help evaluate the situation, collect pertinent information, and provide guidance to the caller.

The propane industry developed a model questionnaire and script to use in situations where customers report propane emergencies such as leaks or releases. The questionnaire prompts personnel answering calls to ask questions such as:

- Where is the leak?
- Do you hear gas escaping?
- Is the leak near any building?
- Is there an odor of gas in the building?

An affirmative answer to these and other questions prompts the gas company operator to read a script that instructs the caller to eliminate ignition sources, evacuate the building to a safe distance, and wait for gas professionals or fire service personnel. Equipping 911 operators with such a prewritten guide can potentially improve safety by initiating important first response actions such as evacuation.

Prepared checklists can also be useful in prompting for information to ask the caller. No one can remember all of the right things to ask. Having a checklist expedites the call taking while still obtaining the information that will benefit the emergency responders. Taking calls for hazardous materials and terrorist incidents can be remarkably different. The use of decision trees and checklists, as previously discussed, can be very beneficial for dispatchers in determining what questions to ask of the caller.

Call takers should get detailed information concerning the location or facility where an incident has occurred. Even if hazardous materials are not involved in an incident, response personnel should know if there are hazardous materials on the scene or facility. The dispatcher has the first and best opportunity to obtain this information through the questions they ask. Even if the presence of hazardous materials cannot be confirmed, remind responders of the potential and to be on the lookout. If possible, ask the caller to wait at a certain location to meet responders. First-hand information from the caller can confirm information forwarded by the dispatcher.

Generally, threat incidents are hoaxes and no bomb, chemical, biological, or nuclear material is found. If anything is found, it turns out to be a benign material or device. Just the same, incidents have occurred involving explosive materials or devices where the device or material was real and an explosion occurred. All threats should be taken seriously and responded to according to local emergency plans. A large number of all terrorist incidents involve explosives or explosive devices. Some of the important procedures for the call taker to follow include listen, be calm and courteous, do not interrupt the caller, obtain as much information as possible, and initiate call trace action. The caller may be the person who placed the bomb, someone who knows the bomber, or a witness to the device.

Gathering information about the caller can help identify that person. It is important that the call taker identify as much information about the caller as possible. Information should include sex; estimated age, accent, voice volume, speech; diction, and manner. Nuclear threats, especially bombs, do not have a lot of credibility. The U.S. Department of Energy (DOE) operates a Nuclear Emergency Search Team (NEST), headquartered at Nellis Air Force Base near Las Vegas, Nevada, who should be called in addition to the FBI for any nuclear threat.

Anthrax threats have become the bomb threat of the 21st century. Threats have included telephone calls, mailed letters and packages, items left in buildings, powder left on building surfaces and air handling systems, and notes written on walls. Anthrax is a biological agent that could be used by terrorists to further their cause. Because anthrax is such a credible threat, all reports should be taken seriously. The response should be within local guidelines for anthrax threats. However, consideration should be given to FBI guidelines, which suggest a low-key response. Every one of the dozens of anthrax threats that have occurred across the country has been a hoax. Nonetheless, response personnel were tied up for hours, businesses were disrupted, traffic was disrupted, citizens underwent decontamination, antibiotics were administered to "victims," and the cost to taxpayers was in the tens of thousands of dollars.

The threat of anthrax is real, but the terrorist does not have to have anthrax to cause disruption of the daily lives of citizens and create fear among the population, and just saying they have anthrax can further the cause of a terrorist. Lessons learned from anthrax hoaxes are not to overreact. Evacuation of large numbers of people may not be necessary if the material is not in an air handling system. Anthrax is not contagious from person to person, so quarantine is not required. All emergency responders should wear respiratory protection. Decontamination is only done for the peace of mind of those near the material. Only soap and water should be used, not chlorine bleaches. Antibiotics should not be administered until or if there is a positive test result for anthrax. Samples of the material should be taken and submitted to a laboratory for testing.

Nearly every fire, EMS, and law enforcement response has the potential for hazardous materials to be present. Terrorism incidents occur on a less frequent basis but can present challenges to the emergency response system nonetheless. Call takers should use their knowledge and training to determine, based upon the caller's information, more specific information about hazardous materials and terrorist agents, which may be present. This information should be passed on to emergency responders to assist them in the mitigation of the incident. Working together, call takers and emergency responders play a critical role in determining the successful outcome of hazardous materials and terrorist emergencies.

Response to Hazardous Materials Incidents

According to Occupational Safety & Health Administration (OSHA) 29 CFR 1910.120 (q), every emergency responder who responds to the scene of a hazardous materials incident must be trained to a minimum of operations level. This requirement has been around since the mid-1980s, however, to this day, there are emergency responders who do not even have hazardous materials awareness training let alone the requirements of operations level. This is the equivalent to sending firefighters to a fire incident without basic training; emergency medical services (EMS) personnel to medical emergencies without medical training; and police officers to an armed robbery without firearms training. OSHA regulations aside, it just makes good sense to prepare emergency responders to deal with hazardous materials. They do not happen often, I grant you, but they should be trained and equipped to do the job they are asked to do. We do not always know an incident has dangerous chemicals involved so all responders need training to be able to recognize potential hazmat incident scenes.

Hazardous materials are everywhere in every size of jurisdiction those with millions of people to Monowi, Nebraska Population 1. We can't legally say we don't need hazardous materials training because we do not deal with hazardous materials. Maybe without training you didn't realize what you were dealing with was a hazardous material. Volume One teaches us that for over a 100 years responders did not respond to what we would call hazmat incidents with any level of formal training. Many responders were killed unnecessarily because of a lack of knowledge of the dangers of the materials they were dealing with. Maybe they knew the dangers but felt they had to do something anyway. Is the 21st century we need to be trained to a minimum of operations level, and need to know what to do and what not to do based upon our level of training and knowledge. There is no getting around it.

Responding to acts of terrorism has also become the target of specialized training for emergency responders. Response to terrorism requires training for first responders in addition to operations training required to respond to hazardous materials incidents. Former Chief of Hazmat John Eversole of the Chicago Fire Department has always said: "if you can't do hazmat, you can't do terrorism." There is a lot of truth in that statement. Therefore, responder training should address terrorism awareness in addition to hazardous materials awareness training.

When the OSHA requirements first surfaced, it was difficult to convince the response community that the training was needed. There were chiefs who declared "we won't respond to hazardous materials incidents, then we won't require the training." Well think about it, how many hazardous materials incidents or other unidentified incidents have the potential for hazardous materials and are called in or dispatched as hazardous materials incidents? Many emergencies are fires, EMS responses, traffic accidents, or crime events, which involve hazardous materials as an additional problem.

Often responders are unaware of what to look for in a potential hazardous materials situation. They may become victims of hazardous materials and not know it until they experience the injury, illness, or death from the exposure. That is why ALL emergency response personnel, fire, EMS, and police should have a minimum of operations-level training. Just as not all hazmat incidents are dispatched as such, not all acts of terrorism will be realized when the first call comes in.

Four major first responder functions upon arrival at the scene of a hazardous materials incident are Recognition, Identification, Notification and Protection. These need to be accomplished to begin incident stabilization and set the foundation for the remainder of the incident actions.

Recognition

Recognition is the ability to determine when hazardous materials are present or might be present because of circumstances or clues at an incident site (Figure 4.10). Recognition is the most important function of the first responder for both hazardous materials and terrorist incidents. If the hazardous materials or terrorist incident is not recognized, then nothing else matters in terms of first responder tasks. Although terrorism involves different circumstances and recognition clues, most operations performed by hazardous materials personnel, including first responders, would be similar.

Figure 4.10 Recognition is the ability to determine when hazardous materials are present.

Clues to Hazardous Materials Presence

Six clues are typically addressed by first responders for identifying hazardous materials: occupancy/location, container shape, markings/color, placards and labels, MSDS and shipping papers. These clues require the use of human senses, including sight, hearing, and smell to indicate the presence of hazardous materials. Responders should not use their sense of smell, but someone on scene may indicate certain smells were present. Additional clues can also be helpful, including odors, dead animals, gas leaks, fire or vapor cloud, damaged vegetation, visible corrosive actions or chemical reactions, dead marine life, pooled liquids, hissing of pressure releases, condensation lines on pressure tanks, injured victims, or casualties.

Occupancy/Location

Hazardous materials are stored and used in every community (Figure 4.11). Responders should be able to identify potential locations of hazardous materials in the community by the type of occupancy and location where they may be stored or used. For example, in many small communities across the country, farm chemical supply stores may contain pesticides, anhydrous ammonia, propane, ammonium nitrate, motor vehicle fuels, and other chemicals. If a farm chemical supply store is located in your community, expect to find these chemicals there. Hospitals may contain liquid oxygen, anesthetic gases, cleaning supplies, and disinfectants, to name a few. Water and sewage treatment plants may contain chlorine and other treatment chemicals. High schools and colleges can contain many different kinds of chemicals in science classrooms, flammable liquids and gases in shop areas, and cleaning supplies. The list can go on and on.

Figure 4.11 Hazardous materials are stored and used in every community.

Responders should conduct preplanning inspections of fixed facilities to determine what chemicals are stored and used there. First responders can also become familiar with occupancies and locations that could become the target of terrorists using hazardous materials. These might include government buildings, places where large numbers of people assemble, public transportation and infrastructure, telecommunications facilities, public utilities, and historical locations.

Modes of Transportation

Transportation incidents make up a large percentage of hazardous materials responses. Knowing how to recognize transportation containers that may carry hazardous materials is not that different from recognizing occupancy/location. It is just that transportation containers can be found anywhere in the transportation system present in our jurisdictions. Transportation modes that are regulated by the U.S. Department of Transportation (DOT) include highway, rail, air, water, and pipeline.

Container Shapes and Sizes

Highway Bulk Containers Although placards on tankers provide information concerning the hazard class of the material inside, a great deal can also be learned about the physical characteristics and hazards of the material from the tank itself. For example, the MC/DOT 312/412 is used exclusively for transporting corrosive materials. The tanker, however, may carry oxidizer, flammable, poison, or corrosive placards.

Placards are placed on a vehicle based on the U.S. DOT system of listing the most severe hazard. Corrosive materials, as with most hazardous materials, can have more than one hazard. If response personnel become familiar with the various types of bulk highway containers and the products they typically haul, they can better identify "hidden" hazards of materials involved in accidents.

The DOT defines a highway bulk cargo tank as "any tank attached to or forming a part of any motor vehicle or any bulk liquid or compressed gas packaging not permanently attached to any motor vehicle, which by reason of its size, construction or attachment to a motor vehicle, is loaded or unloaded without being removed from the vehicle." The term "bulk hazardous material" is defined by the DOT as "any container with over 119 gallons liquid capacity." Other vehicles are also used to transport dry bulk and liquid/dry mixtures.

Specifications for construction of highway transportation bulk tanks are in accordance with the American Society of Mechanical Engineers (ASME) requirements. Additionally, the DOT stipulates procedures for manufacturing, maintaining, testing, inspecting, and repairing bulk containers. Construction materials are also specified by the DOT and include aluminum, carbon steel, high-strength/low-alloy mild steel, and stainless steel. Tanks are designed keeping in mind the hazards of the materials they will be hauling.

DOT regulations and construction standards apply to both interstate and intrastate vehicles. The most common bulk containers used to transport hazardous materials are MC/DOT 306/406, 307/407, 312/412, MC 331, MC 338, dry bulk, and tube trailers (MC stands for motor carrier and DOT stands for Department of Transportation). Since August 31, 1993, the manufacture of vehicles with the MC 300 series specifications is not allowed. Vehicles manufactured after the above-mentioned date are required to conform to design specifications for the DOT 400 series. DOT 400 series tanks are built using stronger metals, are larger, and have a larger volume and a lower center of gravity. MC 300 series tanks will remain in service as long as they meet testing and inspection requirements.

MC/DOT 306–406 tanker (Figure 4.12) is primarily an atmospheric pressure, noninsulated, flammable liquid container that is hydrostatically tested to 3 psi. Capacities vary from 2,000 to 9,000 gallons. It generally has an elliptical shape, although some manufacturers make a round version, and is used to haul gasoline, diesel, aviation fuel, and other flammable liquids. Materials used to construct these tanks include aluminum, steel, and stainless steel. Baffles installed within the container limit product movement during transportation.

These tanks may have multiple compartments possibly carrying several different materials. Specification plates, usually located on the right frame rail of the trailer, are small and difficult to read. They may be of more use during training and preplanning than during an incident. The plate contains information about the type of tank, manufacturer,

Figure 4.12 MC/DOT 306-406 liquid tanker.

construction material, date built, design and test pressures, number of compartments, and capacity.

Specification plates for other bulk tanks contain similar information and are located in the same general area. Relief valves on 306/406 tanks are spring-loaded and remain closed during transportation. Valve operation can be mechanical, pneumatic, or hydraulic. Each valve is equipped with an automatic heat-activated closure system. This system is usually a fusible link but could operate by some other means at temperatures up to 250°F.

There is also a secondary closure system that is separate from the fill and discharge valves and mounted inside the tank. Manual controls for the secondary system are usually on the left front of the container. This type of tank is unloaded on the bottom. Bottom valves are designed to shear off in the event of a collision with such force that the valve would be damaged. There may be up to 10 gallons of fuel in the valve and piping system under the tank. Fill openings on top of the tank are protected by manhole covers that are securely closed.

Tanks are provided with rollover protection and a safety device which prevents the covers from releasing when excessive internal tank pressure exists. Vacuum and relief vents are located on top of the barrel of the tank or internally. The vacuum vent is set to open at 6 oz of vacuum and the relief vents open with as little as 1 psi. Both vents are designed to prevent the release of the product during a rollover. If mounted on the outside of the tank, valves must be protected from rollover damage. Some 306/406 tanks have vapor recovery systems to prevent vapors from reaching the atmosphere. Bumpers extend 6 in. from any vehicle part and serve to protect the vehicle and provide a method of gauging the impact of an accident involving the vehicle. Response personnel can use the information during damage assessment. If the bumper is significantly damaged, the tank's baffles are also likely to have been damaged and their integrity compromised.

MC/DOT 307/407 (Figure 4.13) is a low-pressure tank used for transporting flammable liquids, Class 6.1 poisons (poison liquids), and light corrosives. Working pressure in the tank is 25 psi and not greater than 40 psi. Tank capacities range from 2,000 to 8,000 gallons, and they are constructed using aluminum or stainless steel. This tank may be insulated or uninsulated. Insulated tanks have a horseshoe shape when viewed from the rear. Uninsulated tanks are generally round and some have reinforcing rings similar to the 312/412; however, the tank diameter is much larger. Insulated tanks that no longer meet low-pressure specifications are sometimes used to transport molten solids such as asphalt. These materials have temperatures in excess of 300°F and can be a thermal hazard.

Vehicles carrying molten materials carry a "HOT" placard and sometimes also have a miscellaneous hazardous materials designation. Valve location and information is the same as for the 306/406. Pressure lines may be present on the 307/407, which contains inert gases that may be

Figure 4.13 MC/DOT 307/407 liquid tanker.

injected into the tank to absorb moisture or used to assist in offloading the product. Pressure vents are installed in the 307/407 tanks to limit the internal pressure to 130% of the maximum allowable working pressure. Vents are pressure-activated by a spring-loaded mechanism. Fusible and/or frangible (breakable) venting may be provided with fusible vents activating at temperatures of 250°F. Frangible disks are designed to burst at not less than 130% or more than 150% of the maximum allowable working pressure.

MC 312/DOT 412 tankers (Figure 4.14) are used exclusively to transport corrosive liquids. However, DOT's hierarchy system lists corrosive materials as Class 8. Some corrosive materials may be flammable and oxidizers. These are Class 3 and 6. So even though 412 tankers are exclusively for corrosives the placards may be flammable or oxidizer or something else. Just understand that the tank shape indicates the material is also a corrosive. Corrosives are usually heavy materials, so the tank has a small diameter with reinforcing rings around the outside. These tanks are considered low-pressure containers much like the 307/407. The 312/412 tanks are constructed of black iron steel, stainless steel, or an aluminum alloy, with a lining that will resist degradation and reaction with the tank's contents. Unlike the previously mentioned tanks, these tanks are loaded and unloaded from the top. No valves or piping can be found on the underside of the tank. Valves and venting systems are much the same as the previously mentioned tanks. The 307/407 tanks can be insulated or uninsulated. They may have a steam pad that can be used to heat the container's contents and speed the offloading process. It is located at the rear of the tank.

Figure 4.14 MC 312/DOT 412 corrosive tanker.

MC 331 tanks (Figure 4.15) are high-pressure containers used to haul liquefied gases including, propane, liquefied petroleum gas, chlorine, and anhydrous ammonia. Liquefied gases are gases that have been liquefied by bringing the gas to its critical temperature and pressure. At this point, the gas turns into a liquid. Propane, for example, has a critical temperature of 206°F and a critical pressure of 617 psi. At that temperature and pressure, the propane gas becomes a liquid. Butane has a critical temperature of 306°F and a critical pressure of 555 psi. Once liquefied, the gas is kept in the liquid state by pressurizing the tank to maintain a constant

Figure 4.15 MC 331 liquefied compressed gas tanker.

artificial atmosphere pressing down on the liquid to keep it from returning to the gas state.

Liquids in these tanks are at atmospheric temperature. Whatever the temperature is outside, the liquid inside will be nearly the same. Liquefied gases exist in the tanks well above their boiling points. The pressure in the container keeps the material from boiling and turning back to a gas. These tanks are usually constructed using steel with working pressures generally between 100 and 500 psi. Propane has a working pressure of 250 psi, while anhydrous ammonia has a pressure of 265 psi.

These tanks are never filled to the top. There is usually a 20% vapor space allowed above the liquid level. Usually, these tanks are not insulated but rather painted with white or aluminum paint to reflect radiant heat. When a pressure container such as the 331, with a boiling liquid inside, is exposed to flame on the vapor space, the container will quickly fail and produce a boiling liquid expanding vapor explosion (BLEVE). During a BLEVE, when the container opens up, all of the liquid inside instantly turns into a gas because it is existing above its boiling point. If there is an ignition source present, a fireball may also be created.

According to the National Fire Protection Association (NFPA), a BLEVE will occur within 8–30 minutes after flame impingement starts. On average, 58% of BLEVEs occur in 15 minutes or less. If flame impingement occurs on the liquid level of the tank, the liquid in the tank will absorb the heat and protect the container. The heat from the fire will cause an increase in the pressure in the tank which will be vented through the relief valve.

If the relief valve cannot relieve the pressure as fast as it builds up, the container may still fail. Relief valves are designed only to relieve pressures created by increase in ambient temperature, not flame impingement. Excess flow valves are installed at product discharge openings. They operate in the event of a failure in discharge hoses or piping. Venting systems are either mechanical pressure relief or frangible disks. Valve protection during a rollover is much the same as for other MC series tanks.

MC 338 tanks (Figure 4.16) are used for transporting cryogenic gases, sometimes referred to as refrigerated liquids. These materials are very cold with boiling points of –130°F for carbon dioxide to –452°F for liquid helium. Common cryogenics include oxygen, nitrogen, helium, argon, and others. Many of the materials carried in 338 tanks are considered inert gases. That is, they do not readily react chemically to other materials, are not flammable, and are not poisonous. They do, however, have significant hazards when released as liquids or gases. Liquids are extremely cold and can cause frostbite and solidification of anything they contact, including body parts. Liquids also have large expansion ratios, producing huge amounts of vapor from the small spill. In some cases, as little as 1 gallon of a cryogenic liquid can produce over 900 gallons of

Figure 4.16 MC 338 cryogenic tanker.

gas. While these gases are inert in many cases, they can still displace the oxygen in air and cause simple asphyxiation.

Great care should be taken when dealing with liquid or vapor leaks from an MC 338 container, which actually is a tank within a tank. Between the inner and outer tanks there is a space for insulation to help keep the material within the tank from heating and vaporizing. The tanks are not refrigerated and even with the insulation a certain amount will vaporize, increasing the pressure in the tank and causing the excess pressure to be exhausted through the relief valve. This is a normal process and does not indicate a leak in the tank.

Cryogenic tanks are constructed of steel on the outer tank and special steel alloys on the inside to withstand extremely cold temperatures and internal pressures of 23.5–500 psi.

Some additional types of bulk containers warrant discussion because of the dangers of the products inside. Dry bulk tanks carry dry materials of very fine particle size. These materials rarely are placarded and are not regulated by the DOT. However, if released in air in the presence of an ignition source, they could produce a dust explosion. The only dry bulk tanks I have ever seen placarded were those carrying ammonium nitrate, with an oxidizer placard and those carrying ammonium nitrate fuel oil mixture, which is placarded as a blasting agent/Class 1.5 explosive.

Tube trailers (Figure 4.17) are used to transport high-pressure gases. Unlike the MC 331, there are no liquids in these containers, and therefore, no protection against flame impingement anywhere on the surface of the tank. Pressures in these tanks may be in excess of 3,000 psi. These trailers are actually a series of small pressure tanks placed on a flatbed trailer, and

Figure 4.17 Tube compressed gas trailer.

then banded and cascaded together. Accidents have been reported where the tanks have come loose during an accident and come through the cab of the truck.

Another type of truck used for hazardous materials is the **blasting agent mixer** (Figure 4.18). This truck looks very much like an agricultural feed truck. It has three tanks – one for dry ammonium nitrate, one for fuel oil, and one for a mixture of the two. Once mixed, the blasting agent is offloaded through a mechanical arm, much like grain from a feed truck. This vehicle usually carries three placards – oxidizer, flammable liquid, and explosive 1.5, also known as a blasting agent. Knowing the design

Figure 4.18 Blasting agent mixer.

characteristics of hazmat containers can assist response personnel in safely dealing with a hazardous materials incident involving a bulk container. They can more effectively identify the dangers of the hazardous material as well as provide effective damage assessment on the container.

Railcars

What is a Unit Train?

A unit train is a train made up of railcars transporting the same commodity throughout the entire train. Unit trains typically move from one point of origin (e.g. a shipper's plant) to a single destination (Figure 4.19).

These trains provide customers with efficient and economical service by bypassing normal freight rail classification practices that include separating the cars in a train at a rail yard and repositioning them into other trains.

Examples of unit train commodities are:

- Petroleum crude oil
- Ethanol
- Anhydrous ammonia
- Molten sulfur
- Crude oil
- Coal

Railroad cars that are used to transport hazardous materials are, in many cases, similar to their highway counterparts, but they have much larger capacities. The primary types of railcars are box, hopper, bulk, tank, flat, cryogenic and tube. Types of hazardous materials carried by rail are much

Figure 4.19 A unit train is a train made up of railcars transporting the same commodity throughout the entire train.

like those transported on highways, but again a primary difference is the increased amount of product being transported in individual containers. Additionally, there may be more than one tank car of the same product or multiple products.

Highway emergencies usually involve single amounts or small amounts of several chemicals, but rail incidents can involve from 1 to 100 or more tanks of the same or varying types of hazardous materials. Increased amounts of hazardous materials only compound the situation, making the work of response personnel difficult.

Railroad tank cars contain bulk quantities of hazardous materials and are the primary concern of emergency responders in derailments or other types of rail accidents. In fact, over 80% of all hazardous materials are transported by rail and tank cars accounting for 70% of the total. Most serious railroad hazmat incidents involve tank cars; therefore, they will be the focus of most of this text.

Cargoes Vary Greatly

More than 200,000 tank cars are used in rail transportation in the United States. Most tank cars are not owned by railroads but by private companies. Over 1,200 are owned by the U.S. government and supervised by the military. Government cars are used to transport jet fuels, nitrogen tetroxide (a rocket fuel), and materials used in weapons production. The amounts of hazardous material in a tank car can range from a few hundred gallons to as much as 34,500 gallons. (No tank car is considered empty unless it has been purged of the product and cleaned.)

Railroad tank cars are assigned identification numbers that identify the type of tank – pressure, nonpressure, cryogenic, or miscellaneous. Tank cars transporting hazardous materials can be identified through 100-series numbers regulated by the U.S. DOT. Tank cars that carry nonhazardous commodities such as corn syrup and cooking oil bear 200-series numbers and are regulated by the Association of American Railroads (AAR) are not considered hazardous materials containers.

Numerical designations for tank car designs are as follows:

- Pressure tank car designations – DOT 105, 109, 112, 114, and 120.
- Nonpressure tank car designations – DOT 103, 104, 111, and 115; and AAR 201, 203, 206, and 211.
- Cryogenic tank car designations – DOT 113; AAR 204, and 204XT.
- Miscellaneous tank car designations – DOT 106, 107, and 110; and AAR 207 and 208.

While the numbering system for railroad cars identifies particular types of tanks, the numbers may not be readily visible to emergency responders from a safe distance.

Figure 4.20 On a pressure car all valves and piping are enclosed within a dome at the top center of the tank car.

The most important concern for emergency responders about a tank car is whether it is a pressure or a nonpressure tank. Pressure tanks present a much more difficult and dangerous situation for response personnel than nonpressure cars.

Pressure Tank Cars (Compressed and Liquefied Gases)

Pressure tank cars and nonpressure tank cars can be differentiated by looking at the locations of valves and other piping on the tops of the cars. On a pressure car all valves and piping are enclosed within a dome at the top center of the tank car (Figure 4.20). The dome is designed to prevent damage during a derailment. Nonpressure cars can have unprotected valves and piping on tops of the tanks or, in some cases, on top of the domes. Nonpressure tanks may also have bottom fittings and washouts. Pressure tanks are top-loaded through the dome assembly and are generally used to transport flammable and nonflammable gases and poison gases. The tanks may be insulated or noninsulated and are hydrostatically tested for pressures of 100–600 psi. Pressure tank cars present a much more significant hazard to response personnel than nonpressure cars under heat or flame impingement conditions.

Nonpressure Tank Cars (Liquids)

Nonpressure tank cars are hydrostatically tested from 35 to 100 psi. These tanks do not have all valves inside a single compartment (Figure 4.21). They are commonly used to transport flammable and combustible liquids, flammable solids, oxidizers, organic peroxides, poison liquids, and

Figure 4.21 Nonpressure cars do not have all valves inside a single compartment.

corrosives. (There are exceptions; for example, because of the number of accidents involving ethanol and curde oil cars, the cars have been redesigned to protect valves on top of the containers.) Tank cars are constructed from a variety of materials, including carbon steel, aluminum, stainless steel, nickel, chromium, and iron. Single thicknesses of tank materials range from one-eighth to three-quarters of an inch. Tank car design standards are found in 49 CFR, Part 179 of the DOT Hazardous Materials Regulations. Modifications may occur to tanks that accommodate products transported because of product temperature, flammability, or chemical reactivity. Tanks may be insulated externally to protect against the effects of ambient temperatures. Insulation materials can include fiberglass, polyurethane, or pearlite; cork is used in some older tank cars.

Tanks may also be provided with thermal protection designed to keep tank temperatures below 800°F during a 100-min pool-fire or 30-min torch-impingement test. This protection is provided by a layer of wool or ceramic fiber covered by a one-eighth-inch steel jacket. Thermal protection can also be provided through a textured coating sprayed onto a tank's outer surface. Heat from a flame exposure is absorbed by the coating material and is not transferred to the tank metal.

Tank cars are protected from physical damage during an accident by AAR Type E double-shelf couplers (Figure 4.22) and head shields. Shelf couplers are designed to stay together so they do not puncture another car during a derailment or other accident. Head shields provide an extra layer of metal to help prevent dents and punctures when a pressure tank is hit by another car or object.

Figure 4.22 Shelf couplers protect couplings from coming apart during a derailment and damaging another tank.

During the 1950s and 1960s, there were several derailments that involved propane and butane, also known as LPG (Figure 4.23). These derailments were catastrophic, killing emergency responders and civilians and causing millions of dollars in property damage. Following the changes to tank cars, with insulation, head shields, and shelf couplers, these types of disasters have been reduced significantly.

Figure 4.23 During the 1950s and 1960s there were several derailments that involved propane and butane, also known as LPG. (Courtesy: Waverly, TN Fire Department.)

Cryogenic Tanks

Cryogenic railroad tank cars are usually constructed using nickel or stainless steel as a tank within a tank (Figure 4.24). Cryogenic tanks may also be found within boxcars. Because cryogenic liquids are very cold, insulation is placed between the two tanks and a vacuum pulled on the space to maintain the temperature. This process will allow the tank car a 30-day holding time. Cryogenic cars transport various gases, including flammable hydrogen, liquid oxygen, and poisons. Some cryogenic gases, such as nitrogen and argon, are considered inert. Temperatures of these liquefied gases can range from the warmest, carbon dioxide at –130°F, to the coldest, helium at –452°F. Thermal hazards of these materials are significant.

In addition to thermal hazards, cryogenic liquids have a large liquid-to-vapor expansion ratio. A small leak from a valve or container can create a large vapor cloud. Some ratios are as great as 900 to 1, implying that 1 gallon of cryogenic liquid can produce over 900 gallons of gas.

Multi-unit railcars are used to transport individual tanks of gases in uninsulated "ton containers." They are removed from the car for use, refilled, and re-transported. Products carried include chlorine, phosgene, anhydrous ammonia, and refrigerants. "Ton containers" have a 180- to 320-gallon water capacity and are pressure tested from 500 to 1,000 psi. They may be transported by rail or truck and can be found on special flat cars, boxcars, or gondola cars, as well as on "trailer on flat car" (TOFC) or "container on flat car" (COFC) units.

High-pressure tank cars (similar to highway tube trailers) are approximately 40 ft long and contain a series of 25–30 steel cylinders or individual tanks that are tested to 4,000 psi. High-pressure tanks are not

Figure 4.24 Cryogenic railroad tank cars are usually constructed of nickel or stainless steel as a tank within a tank.

insulated and are equipped with pressure relief valves, and are usually used to transport helium or hydrogen. Another type of pressure tank car is only pressurized during unloading. This car is a pneumatically unloaded, covered hopper car. Pressure is applied during the unloading process and the tank is tested between 20 and 80 psi. This type of tank is used for dry caustic soda.

Warnings Posted

Railroad tank cars, as previously mentioned, are built to DOT specifications for hazardous materials transportation. Certain markings are required to be stenciled on each tank car as a part of the specification requirements. Of particular interest to emergency responders is the requirement that names of certain commodities be stenciled on both sides of the tank car in 4-inch-high letters. Around 50 materials require name stenciling, including anhydrous ammonia, chlorine, and LPG. Additionally, if a material is an "inhalation hazard," that must also be stenciled on the container above the commodity name.

A tank containing hazardous materials will bear a DOT specification including DOT followed by the tank car type, such as 111 or DOT-111. Next to the tank car type will be a letter designating the type of protection the pressure car has for accidents or flame exposure:

- A – the tank has top and bottom shelf couplers.
- S – the tank has an A-plus head puncture resistance.
- J – the tank has A and S and jacket thermal protection.
- T – the tank has A, S, J, and spray-on thermal protection.

If the tank is a pressure container, following the letters will be a number indicating the tank test pressure. Following the test pressure will be letters designating the material used in the tank's construction: Al designates aluminum, N nickel, and C, D, and E stainless steel. Nonpressure tanks will not have the test pressure information. Unless they are carrying certain poisonous gases, such as hydrocyanic acid, pressure tanks have relief valves designed to relieve excess pressure caused by increases in ambient temperature. These relief valves are not designed to relieve the pressure created from radiant heat from a fire or other source or direct flame impingement.

Boxcar Shipments

Boxcars are also used to transport hazardous materials. Individual container sizes inside boxcars are 119 gallons or less, and in many cases, 55 gallons or less. Potentially, any class of hazardous material may be included in boxcar shipments. Flatcars are used to ship pallets of hazardous materials, including small containers. Flatcars are also used to carry "intermodal containers," which can be box containers or any other type of bulk tank container. Intermodal containers get their name from the fact that they are shipped by highway, rail, and water. They can carry any class of hazardous material and the quantities will be smaller than ordinary highway or rail containers. There can, however, be multiple intermodal containers that can present a quantity problem during an accident.

Hopper Cars

Hopper cars do not always contain hazardous materials, but the physical state of the materials in the container may present a hazard. When suspended in air, fine powder and dust can become a dust explosion hazard. An accident could cause the materials to be airborne; if an ignition source is present, a dust explosion can occur. Response personnel who have railroads that transport hazardous materials through their jurisdictions should be familiar with the types of railcars and the hazardous materials they contain.

Intermodal Containers

One of the fastest-growing areas of hazardous materials transportation is that of intermodal containers (Figure 4.25). The use of intermodal containers is referred to by the DOT as "intermodalism," which means "the use of more than one form of transportation." Intermodal containers resemble many of the other bulk transportation containers found in highway, rail, and water modes of transportation, but are smaller in capacity. They are designed to be transported in rail, water, and highway modes without the need to offload the products; that is, the container is transferred from one

Figure 4.25 One of the fastest-growing areas of hazardous materials transportation is that of intermodal containers.

mode of transportation to another, rather than the product being transferred from one container to another.

When aboard a ship, intermodal containers may be found above or below the deck. Once a container ship reaches the port, the intermodal containers are unloaded by giant cranes. The containers may also be shipped on inland waterways aboard barges. Containers are moved around on land by specially designed forklift-type vehicles that are much larger than a typical forklift. Containers may be found stacked on top of each other on the dock awaiting further shipment on land. From the dock, the containers are moved inland by rail or truck. Intermodal containers are not usually transferred directly from ship to rail or truck. They are unloaded onto the dock and then transferred to rail or truck by specially designed cranes. An added advantage of intermodal containers is that they are portable and can be taken to the end-use site and offloaded as a stationary storage tank until the product is used up and then returned for refilling. Intermodal containers are regulated by the DOT, and containers are made to specifications prepared by the International Maritime Organization (IMO), an agency of the United Nations that deals with treaties for maritime safety among other matters. Primarily, IMO Specification 1, 2, 5, and 7 tank containers are used to transport hazardous materials. Intermodal containers come in many sizes and shapes. Length of box containers can vary from 10 to 48 ft. Typical heights are 8 ft, 6 in., but some styles may vary from 8 to 9 ft, 6 in. The standard width of intermodal containers is 8 ft. Containers may have doors at the rear with some also having side doors.

Intermodal containers may be insulated or uninsulated and may have environmental temperature controls. Bulk tanks are designed with a steel structure around the outside to facilitate stacking and movement. They are usually 20–28 ft in length and are used for liquids and bulk materials. Tank containers have a capacity that ranges from 4,000 to 6,000 gallons. They are used for transport of liquids or gases, including flammable, toxic, and corrosive chemicals; cryogenic liquids; and others. Tanks are constructed using metal with two basic components – the tank itself and an outer framework. Usually, containers are constructed of stainless steel, but they can also be aluminum, mild steel, or magnesium alloy. Containers may be lined, refrigerated, heated with electricity or steam, and insulated with metal or plastic jackets. Weight, volume, and construction details of a tank container vary considerably due to the properties of the transported substance.

The IMO defines five different types of tank containers where the following two types are significant for chemicals. Within the United States, IM 101 and IM 102 portable tanks are the most commonly used containers for both hazardous and nonhazardous materials. They are equivalent to IMO 1 and 2 tanks. Specification 51 tanks in the United States are equivalent to IMO 5 tanks and are pressurized between 100 and 500 psi.

IMO 1 tanks for the highly flammable, toxic, and corrosive liquids have the following specifications:

- Shell constructed of 316 stainless steel
- Capacity of 3,158–6,974 gallons
- MAWP of 58 psi (25.4–100 psi U.S.)
- For transport of hazardous liquids
- Available in 7.65-, 10-, 20-, or 30-ft sizes
- Can be heated using steam or electricity (glycol cooling optional)
- Flashpoints below 32°F

IMO 2 tanks for medium-hazard products such as flammable liquids, herbicides, resins, and insecticides have the following specifications:

- Shell constructed of 316 stainless steel
- Capacity of 3,158–6,974 gallons
- MAWP of approximately 25 psi (14.5–25.4 psi U.S.)
- For transport of low-hazard liquids
- Available in 7.65-, 10-, 20-, and 30-ft sizes
- Heating by steam or electricity and cooling by glycol
- Flashpoints between 32°F and 142°F

IMO 5 tanks for high-hazard products such as flammable gases have the following specifications:

- Shell constructed from 316 stainless steel
- Capacity of 3,846–6,710 gallons
- For transport of gases, including propane, butane, anhydrous ammonia, and refrigerant
- Available in 20-, 30-, or 40-ft sizes
- Cooling by sun-shield

IMO 7 tanks for cryogenic/refrigerated liquids have the following specifications:

- Capacity of 2,105–5,263 gallons
- For transport of bulk liquids, including oxygen, nitrogen, carbon dioxide, argon, and ethylene

Some intermodal containers are specially designed to carry specific hazardous materials. These include the Type 7 for cryogenic liquids, sometimes referred to as refrigerated liquids, and tube modules used for compressed gases. Tube modules are used for helium, nitrogen, oxygen, and others. Pressures range from 3,000 to 5,000 psi.

Intermodal containers are marked with placards in the same manner as their full-size highway and railroad counterparts. The only exception is that they are placarded according to highway placard regulations rather than the railroad. On the railroad, all railcars are required to be placarded regardless of the amount of product, so even if they are being shipped by rail, intermodal containers will be placarded according to the DOT's Table 1 and Table 2 placard requirements (FEMA/NFA).

In addition, because these are bulk containers, a UN 4-digit identification number will be assigned to the product and displayed at the center of any placard present or in an orange rectangle on the container. Additionally, international shipments have markings associated with the agreement concerning the International Carriage of Dangerous Goods by Road (ADR). An additional orange placard will be on the container above the 4-number with the ADR Hazard Identification Number (HIN). This is also known as "The Kemler Code."

The ADR Hazard Identification Number HIN (The Kemler Code) The ADR Hazard Identification Number HIN, also known as the Kemler Code, is carried on placards on tank cars and tank containers running by road under international ADR regulations.

The following identification system is in use for placards in Europe (ADR) for road transport. The background of the placard is orange. The border, horizontal line, and figures are black.

X338
1717

33
1088

Identification numbers are shown in such a way, that the upper number is indicating the danger and the lower number identifies the substances with the 4-digit number given in the UN Recommendations on the Transport of Dangerous Goods.

An orange blank placard without any numbers indicates vehicle carrying dangerous load (drums, packages, etc.) or multi-load tanker.

The ADR Hazard Identification Number HIN consists of two or three figures. The first figure of the Kemler Code indicates the primary hazard:

- 2 Emission of gas due to pressure or chemical reaction
- 3 Flammability of liquids (vapours) and gases or self-heating liquid
- 4 Flammability of solids or self-heating solid
- 5 Oxidizing (fire-intensifying) effect
- 6 Toxicity
- 7 Radioactivity
- 8 Corrosivity
- 9 Risk of spontaneous violent reaction

The second and third figure generally indicate secondary hazards:

- 1 the hazard is adequately described by the first figure
- 2 (flammable) gas may be given off
- 3 fire risk
- 4 fire risk
- 5 oxidizing risk
- 6 toxic risk
- 8 corrosive risk
- 9 risk of spontaneous, violent reaction

Doubling of a figure indicates an intensification of that particular hazard. Where the hazard associated with a substance can be adequately indicated by a single figure, this is followed by a zero.

If a hazard identification number is prefixed by letter 'X', this indicates that the substance will react dangerously with water.

Shown on the right is the combination 4-digit number and the HIN for acetyl chloride. The number 1717 is the 4-digit identification number and the X338 is the HIN. The first figure in The Kemler Code indicates the primary hazard (see chart above). In this case, the number is 3 that indicates a flammable liquid as being the primary hazard. The second two numbers are the secondary hazards also taken from the chart for number combinations (on the right).

When a number is doubled, it indicates that the hazard is intensified for that particular hazard. Because there are two 3s, it is an indication that there is an additional fire hazard, which in this case is a self-heating material. The number 38 indicates that this material is a "self-heating liquid, corrosive." This particular chemical also has an X in front of the numbers, which indicates that this material is water-reactive. Therefore, the X338 above the 4-digit number tells you that acetyl chloride is a flammable liquid that is self-heating and it is also corrosive and water-reactive. (Information concerning the ADR HIN can be found in the white pages at the front of the emergency response guidebook (ERG).

Box-type intermodal containers are also a concern to national security as they may be used by terrorists to smuggle personnel and or weapons of mass destruction into the country. Security at ports is considered to be one of our vulnerable points. Congress is working with the Department of Homeland Security and the Coast Guard to beef up security at the nation's ports. People who work at port facilities and those who operate pleasure boats in the vicinity should be alert to unusual activity, which may be associated with potential terrorist actions. The Coast Guard should be notified if anything unusual is observed (DOT ERG).

Ships and Barges

Ocean-going vessels pay call at many ports on the east, west, and gulf coasts, some are tankers and some container ships (Figure 4.26). Containers often contain hazardous materials and are placarded when the intermodal containers are loaded onto the docks, awaiting transfer to rail or truck for continued transportation across the country. Barges carrying bulk hazardous materials travel up and down the nations inland waterways, like the Mississippi, Missouri, and Ohio river systems (Figure 4.27). The cities that have sea or inland ports are generally prepared to deal with hazardous materials. Places like New Orleans, Memphis, St. Louis, Galveston, Corpus Christi, Houston, Los Angles, San Diego, New York, and Norfolk to name a few have hazardous materials teams ready to respond to these emergencies.

Figure 4.26 Ocean-going vessels pay a call at many ports on the east, west, and gulf coasts, some are tankers and some container ships.

Figure 4.27 Barges carrying bulk hazardous materials travel up and down the nations inland waterways like the Mississippi, Missouri, and Ohio river systems.

Pipelines

Pipelines are a regulated mode of transportation by the DOT (Figure 4.28). For the most part, pipelines are underground and not seen by emergency responders. Natural gas pipelines that go into cities develop leaks naturally or by contractor, resulting in fires and explosions that fire departments respond to on a fairly regular basis. Major pipeline emergencies are rare compared to other types of transportation. The ERG has a good section for dealing with pipeline emergencies. Pipelines are only marked regarding the location where they cross another form of transportation, such as a highway or railroad. At these points, there are pipeline markers that identify the type of hazardous material and contact information for the pipeline company.

Fixed Facilities

The U.S. DOT regulates the design and construction of shipping containers to help assure the safe transportation of hazardous materials. Specifications for fixed-facility containers, however, are quite different. There is no mandatory regulation of fixed-facility containers. Recommendations for fixed tank specifications are issued by codes and standards organizations. I have seen a railroad tank car being used as an underground storage tank for a gasoline service station and highway transportation containers being used for fixed storage after they were no longer certified for highway use by the DOT.

Fixed-facility containers can be of almost any size and shape. Some of them have names based on their designs and functions, such as the

Figure 4.28 Pipelines are a regulated mode of transportation by the DOT.

open floating-roof tank. Others, however, do not have any specific name or designation, unlike their highway counterparts. Several NFPA standards address the storage of hazardous materials in fixed containers. These include NFPA 30 Flammable & Combustible Liquid Code and NFPA 58 Storage and Handling of Liquefied Petroleum Gases. Both of these codes address the design and safety requirements for specific fixed tanks such as propane and flammable liquid storage.

The American Petroleum Institute, the Institute of Petroleum, the American National Standards Institute, and others also publish standards, but these are "consensus standards" and do not become law or mandatory until a jurisdiction adopts them. Some jurisdictions have passed local codes and ordinances that regulate certain hazardous materials in fixed containers; however, these vary from location to location and do not have the consistency that the DOT regulations provide for the transportation of hazardous materials in containers.

Bulk Petroleum Storage

The first tanks to be discussed are those used for bulk petroleum storage. These include tanks with cone roofs, floating roofs, open floating roofs, and retrofitted floating roofs. Many of these tanks are associated with tank storage facilities or "farms" where there are multiple tanks at one facility. Tank farms are often connected with pipelines as well as highway, rail, and waterway transportation.

Fixed storage tanks have pressures that range from atmospheric (0–5 psi) and low (5–100 psi) to high (100–3,000 psi) and ultra-high (above 3,000 psi). Bulk petroleum tanks are generally considered atmospheric pressure tanks.

Figure 4.29 Cone-roof tank.

Cone-roof tanks (Figure 4.29) get their name from the inverted-cone-shaped construction of their roofs. They are atmospheric and low-pressure tanks with cylindrical outer walls supporting the cone roof. American Petroleum Institute Specification 650 calls for the roof-to-shell seam for this tank to be designed to fail as the result of a fire or explosion, reducing the possibility of a pressure buildup. Cone-roof tanks may be used to store gasoline, fuel oil, diesel fuel, and corrosive liquids. Some of these chemicals, such as gasoline, are very volatile and easily produce vapor at normal atmospheric temperatures. Because a cone-roof tank is open inside, the surface of the liquid is exposed to air and product will be lost to vaporization. Therefore, a cone-roof tank is primarily used for nonvolatile materials. Contents of cone-roof tanks change frequently: daily, even hourly under some circumstances. These tanks are recognized by the cone-shaped roof and the lack of wind girders or external vents around the top of the sidewall.

Open floating-roof tank (Figure 4.30) has a roof that literally floats on the liquid product in the tank. The outer walls of these tanks are vertical and cylindrical, with the floating roofs eliminating the vapor space in the tank. Drains in place are designed to remove moisture accumulation on the surface of the tank. An articulating-arm-type device on the top of the tank functions as a ladder for examining the interior. As the product level and floating roof go down, so does the end of the arm, which rests on top of the roof. The location of this arm, when viewed from the outside, can be an indicator of the approximate amount of product in the tank. If the arm is completely visible, the tank is near full. If the arm is completely inside the tank, it is near empty. A wind girder is located near the top and on the outside of the tank. Primarily a retaining band, the wind girder provides

Figure 4.30 Open floating-roof tank.

Figure 4.31 Covered or internal floating roof.

necessary rigidity to the container wall when liquid levels are lowered. An external staircase is also located on the tank wall to provide access to the articulating arm (walkway) on top of the floating roof.

Covered or Internal Floating Roof (Figure 4.31). These tanks may resemble cone-roof tanks but are distinguishable by the vents around the tank near the roof to sidewall seam. The cone roof provides protection from rain and snow, while the internal floating roof eliminates the vapor space above the liquid and helps prevent the loss of vapor into the air. This style of tank is sometimes used for polar solvent materials, which are miscible with water. If water or snow enters the tank, it would dilute the product. Materials stored in the internal floating-roof tank are generally flammable and combustible liquids because of the vapor loss protection of the floating roof on top of the liquid. A floating-roof tank may also be covered by a geodesic dome that keeps outside weather from affecting the floating roof in the tank. This form of internal floating-roof tank is also used for flammable and combustible liquids.

Horizontal Tanks Horizontal tanks are another kind of tank used for chemical storage in many parts of the country, particularly in rural areas (Figure 4.32). These tanks range in size from hundreds of gallons to thousands of gallons with low or atmospheric pressure. The ends of the tanks are flat, which usually indicates a limited amount of pressure. Many horizontal tanks are supported on stands constructed of metal. Unprotected metal can be dangerous during fire conditions as the metal will be stressed by the heat and fail very quickly.

In the 1950s, six Kansas City, KS, firefighters were killed when the steel supports of a horizontal tank at a gasoline service station failed, sending

Figure 4.32 Horizontal tanks flammable liquids.

Figure 4.33 In the 1950s, six Kansas City, KS firefighters were killed when the steel supports of a horizontal tank at a gasoline service station failed, sending burning gasoline into their location on the street. (Courtesy: Kansas City, MO Fire Department.)

burning gasoline into their location on the street (Figure 4.33). NFPA 30, Flammable and Combustible Liquids Code, requires the steel supports of horizontal tanks be protected from flame impingement by encasement in concrete or other nonflammable material. Horizontal tanks may be used to store flammable and combustible liquids, corrosive liquids, and many other types of hazardous materials. A direct result of the Kansas City fire was the requirement for gasoline tanks at service stations frequented by the public to have underground storage tanks.

According to the NFPA, there has never been a fire or explosion involving an underground fuel-storage tank. During the 1970s and 1980s, it was discovered that underground storage tanks, which were usually constructed of steel, eventually corrode and leak fuel into the ground. As gasoline and other flammable fuels have a specific gravity less than water, they will float on water. Moisture that accumulates inside underground storage tanks settles to the bottom of the tank and causes the metal to corrode. Because of the leakage, the U.S. Environmental Protection Agency (EPA) requires that all underground tanks be replaced and safeguards put in place to prevent and detect leaks.

An alternative to the placement of tanks underground was allowed in the form of an aboveground vault. The vault is constructed of concrete and the tank(s) is placed within the vault, partially underground with the vault top above the ground. There is access to the vault from aboveground. Vaults for flammable and combustible liquids must have safety precautions such as monitors installed and must be able to hold the contents of the largest tank if it should leak.

High-Pressure Tanks Propane and anhydrous ammonia are stored in high-pressure tanks that contain many safety features that are built-in through codes and standards. Industry associations are also heavily involved in safe storage and handling requirements for their members and emergency responders. The National Propane Gas Association (NPGA), in conjunction with the Propane Education & Research Council, has developed a training program for emergency responders with a 220-page text, *Facilitator's Guide*, CD-ROM, and 50-min video.

Propane tanks range in size from the 5-gallon containers used with barbecue grills and 250-gallon tanks used for home heating to bulk-storage tanks containing thousands of gallons of product at propane facilities (Figure 4.34). Propane tanks generally have rounded ends, which is a primary indicator of a pressure vessel. Pressure tanks are equipped with relief valves to vent excess pressure caused by increases in ambient temperature. Propane tanks have extensions on the relief valves to extend vapors well above the tank in the event the vapors catch fire (Figure 4.35). The normal relief valve height is between 6 and 10 in. This height allows ignited vapors to lie on the vapor space of the tank, causing it to fail quickly. Anhydrous ammonia tanks do not have the extension, so if you see a horizontal pressure tank with tall relief valves, it is likely to contain propane; if the relief valves are near the tank's surface, it is likely to contain anhydrous ammonia.

Anhydrous ammonia, propane, and other petroleum gases are liquefied under pressure before being placed in these tanks (Figure 4.36). The liquid level in the tanks is generally 80% of the tank's capacity to allow for vapor space. Pressure containers can be dangerous under fire conditions.

Figure 4.34 Propane tanks range in size from five gallon to thousands of gallons in large bulk tanks.

Figure 4.35 Propane tanks have extensions on the relief valves to extend vapors well above the tank in the event the vapors catch fire.

Relief valves are designed only to relieve normal pressure increases, and not those caused by flame impingement from a fire or radiant heat sources.

In the past 20 years, firefighters in the United States and Canada have been killed fighting fires involving high-pressure fixed propane containers. Flame impingement on the vapor space (Figure 4.37) of the tank can cause the metal to fail, and a BLEVE can occur within 8–30 min of the start of the flame impingement. On average 58% fail within 15 minutes or less.

***Figure* 4.36** Anhydrous ammonia is liquefied under pressure before being placed in these tanks.

***Figure* 4.37** In the past 20 years, firefighters in the United States and Canada have been killed fighting fires involving high-pressure fixed propane containers. (From Chemical Safety Board.)

The liquid level of the tank will absorb the heat from flame impingement and not fail as long as the liquid is in the tank. The flame impingement on the liquid level causes the propane in the tank to boil faster and produce more vapors. If the vapor is produced faster than the relief valve can vent it, tank failure can occur.

Spiracle tanks are another type of high-pressure tanks that may be found at refineries or other locations. These tanks look like giant balls suspended in a steel support structure. Tanks of this type can be used for propane as well as LPG, natural gas, and hydrogen (Figure 4.38).

Tube Banks Tube banks are a type of ultra-high-pressure tanks that are often found at compressed gas companies and distribution facilities (Figure 4.39). The pressures in this type of tank can be in excess of 3,000 psi. Tube banks are actually a series of individual tanks stacked together with valves and piping into a single outlet/inlet, similar to a fire department cascade system. Unlike liquefied gases in high-pressure tanks, the materials in tube banks are gases. There is no liquid space to absorb heat. Under fire conditions, these tanks can fail very quickly.

Cryogenic Tanks Vertical cryogenic storage tanks are often found next to manufacturing buildings, hospitals, bottled-gas companies, and welding-supply houses (Figure 4.40). A heat exchanger next to the tanks is confirmation that these are indeed cryogenic tanks. Heat exchangers are a series of silver-colored tubes with fins next to the tanks. The heat exchanger allows for the cold liquids to be turned back into gases for use. Ambient air around the tubes and fins of the heat exchanger warms the cold cryogenic liquids into their gas state.

Figure 4.38 Tanks of this type can be used for propane and other LPG, natural gas, and hydrogen.

Figure 4.39 Tube banks are a type of ultra-high-pressure tank that are often found at compressed gas companies and distribution facilities.

Figure 4.40 Vertical cryogenic storage tanks are often found next to manufacturing buildings, hospitals, bottled-gas companies, and welding-supply houses.

Cryogenic tanks have narrow circumferences and are very tall. These are high-pressure tanks used to store cryogenic liquids, which are very cold. Cryogenics have boiling points of –130° to –452°. Some tanks, particularly those found at cryogenic-production facilities, may each hold as much as 400,000 gallons. Types of materials found in cryogenic containers include natural gas, argon, nitrogen, chlorine, and oxygen. They can be flammable, oxidizers, or poisons.

Portable Containers

Hazardous materials containers can be found as many different types of portable containers, which may be in transportation, storage, and use (Figure 4.41). Portable containers are used to hold hazardous materials in small quantities, which are easily moved from one location to another. Portable containers are constructed of many different materials including, glass, aluminum, stainless steel, steel, plastic, wood, cardboard, lead, and others. Portable containers are regulated by the U.S. DOT during transportation and governed by ASME, NFPA, and OSHA standards in fixed storage and use. However, portable containers can be made of any material that will hold the contents and may not be specification containers. In some cases, containers may be reused from other purposes, which may create an unsafe storage condition.

Portable containers include, but are not limited to, the following types, wooden and fiberboard boxes, metal drums, fiberboard drums, plastic pails, glass carboys in protective containers, cylinders, ton cylinders, mailing tubes, special lead containers for radioactive materials, plastic, and multi-wall paper bags. Liquid quantities can range from a few ounces to several hundred gallons. Dry materials may range from a few ounces to several hundred pounds. Gases generally do not have much weight but require a substantial container to withstand high pressures causing the tank to be very heavy.

Fiberboard drums and multi-wall paper bags are used for dry materials such as ammonium nitrate fertilizer and calcium hypochlorite, both of which are oxidizers. If the containers become wet during transportation or storage, or during firefighting operations, the packaging material can

Figure 4.41 Hazardous materials containers can be found as many different types of portable containers which may be in transportation, storage, and use.

become impregnated with the oxidizer once the moisture evaporates. If the packaging materials are then exposed to heat or flame in a fire, they can burn vigorously because of the oxidizer embedded in the container material.

Glass and plastic carboys as well as stainless steel kegs, 55-gallon drums, pint, gallon glass bottles, 5-gallon plastic pails, and lined drums are used to transport and store many different types of acids and bases and a variety of other hazardous materials. Acids can react quickly with their container causing container failure and subsequent material spill if incompatible construction materials are used. Not all acids and bases can be placed in the same types of containers. Damage can occur to a container if different acids or bases are mixed in a container. Plastic and glass bottles, 55-gallon drums, and plastic and metal pails may also be used for flammable, poisonous, or oxidizing liquids. Glass and plastic bottles placed inside lead containers are used for some radioactive isotopes. More substantial lead containers are used for high-level radioactive materials because of the shielding required to keep the radioactivity from leaving the container.

Portable container pressures range from atmospheric in the case of drums and bags to ultra-high pressures of 6,000 psi for cylinders. Nearly every type of hazardous material found in bulk quantities may also be found in small portable containers. Additionally, there are many hazardous materials that are not shipped or stored in bulk quantities, but rather are usually found in small containers, such as certain explosives. Ammonium nitrate (ANFO), dynamite, blasting caps, detonation cord, fireworks, and other explosives are packaged and shipped in cardboard boxes.

Quantities of hazardous materials in portable containers are sometimes so small that the DOT does not consider them a serious hazard and does not mandate placard or labeling. These materials are placed in a special class known at Other Regulated Materials (ORM-D). For example, charcoal lighter would be classified as a flammable liquid in bulk quantities; however, in quart cans for consumer use, it is an ORM-D. Oftentimes, there may be large numbers of small containers in a shipment or in fixed storage, which collectively can create a serious hazard even though the individual container quantity is small.

Cryogenic hazardous materials, those with boiling points below –130°F, are shipped, stored, and used in two types of portable containers, Dewar and cylinder (Figure 4.42). Dewar containers are nonpressurized vacuum-jacketed flasks with a 5–20 L capacity, very much like a thermos bottle. Cylinder containers are insulated with a separate vacuum-jacket and a 100–200 L capacity. Cryogenic containers are usually found at gas bottling plants, college and university research facilities, and private-sector research facilities, and are replacing the oxygen gas cylinder in welding operations in many areas.

Figure 4.42 Cryogenic hazardous materials, those with boiling points below –130°F, are shipped, stored, and used in two types of portable containers, Dewar and cylinder.

Common cryogenic hazardous materials include oxygen, nitrogen, argon, hydrogen, natural gas, helium, and others. Cryogenic liquids have very low boiling points and high liquid-to-vapor expansion ratios. Even a small Dewar or cylinder could produce a significant vapor cloud during a release causing the displacement of oxygen in confined areas. While many of the cryogenic liquids are "inert" or nonhazardous, the vapor can still be an asphyxiation hazard when the oxygen in the air is displaced. In addition, the contents of cryogenic containers are extremely cold with temperatures between –130°F and –456°F below zero. Contact with these liquids could cause frostbite, solidification of body parts, or both.

Cryogenic liquids and liquefied gases, such as propane and butane, do not have the same physical characteristics. While cryogenics are very cold, liquefied gases exist at whatever ambient temperature is surrounding the container. If the ambient temperature is 100°F, then the liquefied gas is also around 100°F. If the ambient temperature is –40°F, then the liquefied gas is also around that same temperature. Portable tanks of liquefied compressed gases can be found in many sizes and shapes. They can contain many different types of gases, but commonly contain propane or other LPG used for home barbecue grills, heating living spaces, as a fuel for motorized equipment, soldering, and others. These containers shouldn't be overfilled. There should be a vapor space when the container is filled, usually around 20% of the volume of the container, to allow for expansion of gas within the container during increases in ambient temperature. While not as extensive as cryogenic liquids, LPG also has large liquid-to-gas expansion ratios in the area of 270 to 1, so even a small container can produce a large amount of flammable gas when released to the atmosphere.

Pressures in LPG containers can range from 15 to 230 psi depending on the ambient temperature around the container. The higher the temperature outside the container, the higher the pressure inside the container will be. Container bursting pressures are generally about four times greater than the working pressures of LPG. Rapid heat buildup can still cause container rupture if the pressure cannot be relieved fast enough. LPG is heavier than air and would be found in basements or other low-lying areas during a release. They have no natural odor and are colorless. An odorizer is added to allow leaks to be detected.

Compressed gas cylinders are also quite common and often used for oxygen, nitrogen, carbon dioxide, fire extinguishing agents, hydrogen, and many others. These containers are constructed of heavy steel and have operating pressures of 3,000–6,000 psi. Just the pressure alone in the container can present a significant physical hazard if the valves are knocked off or the containers are exposed to high heat or direct flame contact. These containers can rocket a great distance and present an impact hazard to building occupants and response personnel. Cylinders in storage should have a protective valve cap in place over the valve and be secured in place or to a cart. Exposed valves can be sheared off if a cylinder is knocked over. Cylinders in use should also be secured in place. Cylinders should never be moved or transported with the regulator in place, and it should be replaced with the protective valve cap. Fire code inspectors should watch for proper storage and use of high-pressure cylinders during inspections. While many compressed gas cylinders are painted, there isn't any reliable color code system to identify the contents from the color of the container.

Ton containers (2,000 lb capacity) are used most often for the storage of chlorine and sulfur dioxide gases (Figure 4.43). They are also used to store military blister agents. These cylinders can be found in water treatment facilities, waste treatment plants, and swimming pools. They are shipped in on specially equipped rail and highway vehicles.

Acetylene is a common flammable gas usually associated with welding operations in conjunction with gaseous or cryogenic oxygen. Acetylene is

Figure 4.43 Ton containers (2,000 lb capacity) are used most often for the storage of chlorine and sulfur dioxide gases.

a highly flammable gas with a wide flammable range (2%–80%), which is quite unstable at elevated pressures above 15 psi. The acetylene tank is a small "chunky" tank specially constructed to contain this highly unstable gas. Inert materials such as fullers earth or lime silica are placed in the acetylene container to absorb acetone, a solvent used to dissolve the acetylene gas and maintain its stability. Acetone keeps the acetylene in suspension preventing accumulation of pockets of high-pressure gas, thus stabilizing the explosive tendencies of the gas. Acetylene tanks should never be stored or transported on their side as acetylene gas can separate from acetone producing a potentially explosive situation.

Because of their relatively small size, portable containers can be found on almost any type of transportation mode including private vehicles. They can also be found in most types of occupancies including residences and garages. The quantities of materials present don't create a high level of risk to the community. However, they certainly can be a hazard to response personnel, especially if we are not looking out for them (FEMA/NFA).

Markings and Colors

NFPA 704 Hazardous Materials Marking System Most everyone involved with hazardous materials response is familiar with the U.S. DOT placard and label system for transporting hazardous materials. The U.S. OSHA has also adopted the DOT system and requires hazardous materials placarded in transportation to continue to be placarded during storage and use until all of the hazardous material is consumed and the container has been purged. Placards and labels must remain in place until the material is used and the container is purged of the residue. There are also other fixed storage and use marking systems, including the NFPA 704 system, OSHA

Hazard Communication Systems, and miscellaneous internal marking systems for occupancies, which store and use hazardous materials.

Work on the development of NFPA 704 started in 1952, which resulted in the establishment of the standard in 1957. This system was one of the first efforts to identify locations where hazardous materials were stored and used but does not apply to transportation (Figure 4.44). DOT did not develop their system for transportation marking for another 20 years. Because 704 is a standard, it is not required or mandated unless a local jurisdiction adopts the standard and makes it law within the community. Usage of 704 varies from community to community. Some facilities may place 704 diamonds on buildings and locations where hazardous materials are stored or used voluntarily. So it is quite possible 704 diamonds could be found even in communities where it has not been adopted or required by law. NFPA 704 is a generic hazardous materials marking system designed to alert emergency responders to the presence of hazardous materials at a particular occupancy, to assist them in evaluating the hazards present, and to help them in planning effective fire and emergency control operations. NFPA 704 does not provide specific information about individual chemicals. The information does not provide chemical names. It is a basic identification system to help first responding emergency personnel to decide whether to evacuate the area or to commence control procedures. It also assists responders with selection of fire fighting tactics, appropriate personal protective equipment (PPE), and emergency procedures. NFPA 704 is intended to provide information to emergency responders about the general hazards of materials that may be inside an occupancy. The hazards on a diamond do not provide information about every hazardous material in a facility. NFPA 704 diamonds list the most severe hazards of the most hazardous chemicals present. They are really only a "stop" sign to warn responders and to allow them to be aware of hazardous materials present. Information provided will not indicate routes of entry for toxic materials, degrees of radioactivity, corrosivity, or other specific chemical information. More information will be needed before mitigation efforts are undertaken.

NFPA 704 "Diamonds" are placed on the outside of buildings near the entrances or addresses on the buildings. They may also be placed on the inside of the building where the hazardous materials are actually stored and used. Exact locations are left up to the Authority Having Jurisdiction (AHJ) but minimum requirements in the standard require the following locations:

- Two exterior walls or enclosures containing a means of access to a building or facility
- Each access to a room or area
- Each principal means of access to an exterior storage area

Figure 4.44 This system was one of the first efforts to identify locations where hazardous materials were stored and used, but does not apply to transportation.

Diamonds used on the outside of buildings are generally a minimum of 15 in. × 15 in. in size. Those used inside the building are a minimum of 10 in. × 10 in. Small-sized diamonds may be found on individual containers. NFPA 704 placards are divided into four colored quadrants or four "mini" diamonds within the larger diamond. The upper quadrant at the "12 o'clock" position is red in color followed by blue at "9 o'clock" to the left, yellow at "3 o'clock" to the right, and the bottom is white at the "6 o'clock" position. These colors designate a specific hazard. Red indicates flammability, blue health, yellow instability, and white special information. Special information located in the white quadrant listed in the 704 standard includes materials that react violently or explosively with water, represented by a "W" with a slash through it, and "OX" for materials with oxidizing properties. The standard lists specific criteria for these symbols to be included on the diamond in the appendix. Other users of the 704 system have added other symbols, which are not a part of the 704 standards and include "RAD" for radioactive, "COR" indicating corrosive, "UD" is unclassified detonable, "4D" Class 4 detonable, "3D" Class 3 detonable, and "3N" nondetonable. The NFPA 704 standard is not intended to speak to the following hazardous situations:

- Occupational exposures
- Explosive and blasting agents, including commercial explosives
- Chemicals whose only hazard is as a chronic health hazard
- Teratogens, mutagens, oncogens, etiologic agents, and other similar hazards

Numbers are placed within the quadrants of the 704 diamond indicating the degree of hazard posed by an individual chemical, or as quoted in the standard, "according to the ease, rate, and quantity of energy release of the material in pure or commercial form." The numbers used to identify the range of hazard are from 0 to 4. Zero indicating no particular hazard and four indicating the most severe hazard or most energy release. Degrees of hazard associated with health, flammability, and instability are determined by criteria outlined in the standard. Some jurisdictions may place a letter "G" in the quadrant with the number to indicate the presence of a compressed gas.

Health hazards are ranked according to the level of toxicity and effects of exposure to response personnel. They are based upon short-term, acute exposure during handling under conditions of spill, fire, or similar emergencies. Short-term exposure ranges from minutes to hours. Acute exposures typically are sudden and severe, and are characterized by rapid absorption of the chemical that is quickly circulated through the body and damages one or more vital organs. Acute effects include severe burns, respiratory failure, coma, death, or irreversible damage to a vital organ.

> ***Author's Notes:*** *The following guidelines are my effort to explain the material in NFPA 704 in terms of protection of response personnel. They are not intended to be recommendations. Materials at the scene of a hazardous materials incident should be thoroughly researched and decisions made by those on scene as to the appropriate level of protection responders should wear based upon the hazards present. Specific details of criteria used to determine hazard numbering are located in the NFPA 704 standard.*

- **4** – Materials that can be lethal if response personnel do not wear proper chemical protective equipment. If gases or skin absorbent gases, liquids or solids, Level A chemical protection would be necessary. Firefighter turnouts would not provide appropriate protection. Some examples of Health Hazard 4 chemicals include chlorine, phosgene, hydrocyanic acid (hydrogen cyanide), hydrogen sulfide, phenol, phosphine, pentaborane, and acrylonitrile (vinyl cyanide).
- **3** – Materials that can cause serious or permanent injury. Level A chemical protection or Level B chemical protection would be appropriate depending on the physical state of the hazardous materials. Firefighter turnouts would not provide appropriate protection. Some examples of Health Hazard 3 include anhydrous ammonia, acetaldehyde, acrylic acid, carbon monoxide, formic acid, pyridine, nitric acid, and para-xylene (p-xylene).
- **2** – Materials that can cause temporary incapacitation or residual injury. Level B or Level C chemical protection would be appropriate

for these chemicals. Firefighter turnouts would likely not provide appropriate levels of protection. Health Hazard 2 chemicals include meta- and ortho-xylene (m-xylene and o-xylene), toluene, styrene, ethyl formate, benzene, 1,1,1-trichloroethane, and vinyl chloride.
- Materials that can cause significant irritation. Level C chemical protection would likely be appropriate protection. Firefighter turnouts may provide some protection, particularly respiratory. Many irritants are actually solid materials and may contaminate personnel. Remember that firefighter turnouts are not classified as chemical protective clothing. Health hazard 1 chemicals include acetone and butane.
- **0** – Materials that would offer no hazard beyond that of ordinary combustible materials. Firefighter turnouts would provide appropriate protection for personnel.

Flammability hazards are based upon a material's susceptibility to burning. Conditions present need to be considered as well as the combustibility characteristics of the fuel. Firefighter turnouts are generally the appropriate protective clothing for flammability hazards. There may, however, be conditions where firefighters even in full turnouts cannot be adequately protected because of the volume of fire or flame impingement on containers. Each situation needs to be thoroughly evaluated based on the flammability hazards present. Some flammable chemicals may also be toxic, and toxicity should be taken into account before emergency personnel are sent into a scene where toxic materials may be present or on fire.

- **4** – Materials that are flammable gases, flammable cryogenics, or liquids with flash points below 73°F and boiling points below 100°F (Class IA liquids). Also included are gases, liquids, and solids that may spontaneously ignite when exposed to air. Examples of Flammable Hazard 4 chemicals include hydrogen, acetylene, vinyl chloride, trichlorosilane, propylene oxide, picric acid, phosphorus, natural gas, hydrogen cyanide, formaldehyde gas, ethylene oxide, carbon disulfide, carbon monoxide, and ethyl ether.
- **3** – Liquids having a flashpoint below 73°F and a boiling point at or above 100°F. Liquids having a flashpoint at or above 73°F and a boiling point at or above 100°F (Class IB and Class IC liquids). Also included are solid materials that because of their physical form can form explosive mixtures when suspended in air. Examples of Flammable Hazard 3 liquids and solids include acetonitrile (methyl cyanide), acrolein, aluminum powder, benzene, gasoline, calcium carbide, methyl isocyanate, and potassium.
- **2** – Liquids having a flashpoint at or above 100°F but below 200°F (Class II and Class IIIA liquids). Solid materials that burn rapidly but do not form explosive mixtures with air. Flashpoint solids that give

off flammable vapors. Examples of Flammability Hazard 2 liquids and solids include glacial acetic acid, cresols, lithium aluminum hydride, nitrobenzene, and phenol.

- **1** – Chemicals that will burn in air when exposed to temperatures above 1,500°F for 5 min. Liquids and solids that have a flashpoint at or above 200°F (Class IIIB liquids). Most ordinary combustible materials. Examples of Flammability Hazard 1 materials include trichloroethylene, polychlorinated biphenyls, phosphorus pentasulfide, paraformaldehyde, methyl bromide, magnesium, and anhydrous ammonia.
- **0** – Materials that generally do not burn under normal circumstances.

Instability hazards (reactivity) addresses the degree of inherent vulnerability of materials to release energy. It applies to materials capable of rapidly releasing energy by themselves through self-reaction or polymerization. It does not deal with water-reactive materials. Organic peroxides need to be evaluated with NFPA 432 Code for the Storage of Organic Peroxide Formulations. Instability does not take into account the unintentional combination of materials, which may occur during a fire or other conditions. During storage, unintentional mixing should be considered to establish appropriate separation or isolation. The degree of instability hazard is meant to indicate to emergency personnel if an area should be evacuated, if firefighting should be conducted from a location of cover, if caution is required when approaching for extinguishment, based upon extinguishing agent, or if a fire can be fought using normal procedures. Ranking of instability hazards is based upon the material's ease, rate, and quantity of energy release.

- **4** – Materials sensitive to shock and heat at normal temperatures and pressures. Can undergo detonation or explosive decomposition at normal temperatures and pressures. Examples of Instability Hazard 4 include ammonium perchlorate, 3-bromopropyne, chlorodinitrobenzenes, fluorine, nitromethane, peracetic acid, and picric acid.
- **3** – Materials sensitive to shock and heat at elevated temperatures and pressures. Subject to detonation or explosive decomposition or explosive reaction, but need a strong initiating source or must be heated under confinement. Examples of Instability Hazard 3 include tetrafluoroethylene, silane, n-propyl nitrate, perchloric acid, nitroethane, hydrogen peroxide >60%, hydrazine, ethylene oxide, chloropicrin, and ammonium nitrate.
- **2** – Chemicals that undergo a violent chemical change at elevated temperatures and pressures. Examples of Instability Hazard 2 chemicals include trichlorosilane, titanium tetrachloride, sulfuric acid, styrene monomer, sodium hydride, potassium, phosphorus trichloride, phosphorus, methyl isocyanate, hydrogen cyanide, ethylene, epichlorohydrin, and aluminum chloride.

- **1** – Materials that in themselves are normally stable, but can become unstable at elevated temperatures and pressures. Examples of Instability Hazard 1 materials include acetic anhydride, allyl alcohol, aluminum powder, ethyl ether, hydrogen chloride, magnesium, nitrobenzene, phosgene, and potassium hydroxide.
- **0** – Materials that in themselves are normally stable, even under fire conditions.

Special Hazards listed in the text of NFPA 704 are limited to water reactive and materials with oxidizing properties, which would require special firefighting techniques. Special hazards are indicated by a symbol located in the quadrant of the diamond at the 6 o'clock position in the white colored section. Materials that react violently or explosively with water are identified by the letter "W" with a line through the center. Materials that possess oxidizing properties are identified by the letters "OX." Examples of chemicals with an **"OX" hazard** include sodium peroxide, potassium peroxide, peracetic acid, oxygen, nitric acid, hydrogen peroxide, chromic acid, bromine, and ammonium nitrate. **Water-reactive** materials include calcium carbide, calcium hypochlorite, fluorine, lithium aluminum hydride, potassium, and sodium.

NFPA 704 is just one of the many tools available for emergency responders when evaluating an incident scene for the presence of hazardous materials. More information needs to be gathered before mitigation tactics are decided. Generally, firefighter turnouts do not provide much protection against hazardous materials, although Self-Contained Breathing Apparatus (SCBA) provides a high level of respiratory protection. A thorough evaluation of the hazards present along with a risk–benefit analysis must take place before PPE and tactics are selected (NFPA).

Notification

Once the incident has been determined to be a hazardous materials incident, the first responder must make the proper identifications. Notification of others is required because first responders generally do not have the training or equipment to mitigate the circumstances surrounding a hazardous materials or terrorist incident. Not every community has a hazardous materials team. Notification should be to the nearest hazardous materials team and all other notifications that are in your local plan for hazardous materials incidents.

Dispatchers as part of their training should know who gets called to respond to a hazardous materials incident. First responders should only have to notify dispatch that a hazmat team is needed along with whatever resources are needed. Larger communities with hazmat teams need to establish a level of hazmat response required for certain circumstances. Often incident levels are used. Hazmat personnel should be involved in developing these response levels and should be thoroughly familiar with them. Dispatch should be notified what level is needed, and they should notify the resources based upon the local level requirements. Responders should not have to itemize what resources are needed to the dispatchers unless a resource is needed beyond the typical resources.

Dispatchers should have quick access information available to them with what resources are dispatched on each level of hazmat incident. Dispatchers may also be able to determine the level of hazmat call that has been reported and initiate the appropriate hazmat response level. Responders need to know who in their jurisdiction and who outside their jurisdiction needs to be notified when an incident occurs. Even if the only call they personally make is to the dispatcher, responders still need to know others who will be needed to respond to the scene. However, if a jurisdiction utilizes hazmat levels of response, the dispatcher can be provided with predetermined notification lists based upon the hazmat level of response. The dispatcher should have the actual contact list and make the contacts. Responders may want to contact the National Response Center (NRC) or CHEMical TRansportation Emergency Center (CHEMTREC) directly to obtain information about the chemical involved in an incident. Contact information on both agencies is located in the current edition of the DOT ERG.

Identification

Identification of potentially hazardous material(s) that cannot be determined by call takers can be accomplished by the first responder, if time allows, and can be done safely. When first responders cannot identify hazardous materials involved in a specific incident, then the identification falls to operations or technician-level personnel. We will talk about many identification tools available to all responders to help identify situations where hazardous materials may be present, and if an incident involving hazardous materials may be a terrorist incident. No matter what level of response training you have, every hazmat responder from awareness level to specialist level should know how to recognize when hazardous materials are present, no matter how they got there (FEMA/NFA).

Hints to Hazardous Materials Presence

All Chemical Names Ending in:

-al -ate -ane -azo -ene -ine -ite -ol -one -oyl
-yde -yl -yne -chlorate hypochlorite nitrate perchlorate chlorite

All Chemical Names that Start with or Include

o, p, N, S, sec, tert, m, or numbers

Any Chemical Name that Includes

acet or acetyl	chlorine or chloro	methyl
acid	cis	mono
acrolein	croton	naphtha
acrylonitrile	cyanide, cyanate (ite)	nitride
alcohol	cyano, cyan	nitrile
aldehyde	di	nitric, nitrous
alkali	ester	nitro
amine	ether	octyl
amide	ethyl	pentyl
amyl	ethylene	propyl
arsenous	ethanoic	oxy
aryl	fluorine	penta
azide	fluoro	peroxide
benzene	heptyl	peroxy
benzyl	hexyl	per
bis	hydride	phosphorous
bromine	hydroxide	phosphoric
bromo	iodine	phosphide
butyl	iodo	solvent
carbide	ketone	spirits
caustic	mercuric	tetra
chromium	mercurous	trans
chromous		vinyl

Railcar Marking Requirements

Marking – a descriptive commodity name, identification number, caution (such as inhalation hazard, elevated temperature material, marine pollutant, fumigant, nonodorized), or tank car specification and qualification dates stencils displayed on hazardous material shipments.

Make sure the markings above are displayed on bulk packages as follows:

INHALATION HAZARD Mark

For a material described on the shipping papers as "POISON (TOXIC) – INHALATION HAZARD" or "INHALATION HAZARD," the words INHALATION HAZARD must appear (in at least 3.9-in. high letters) on both sides of the railcar, trailer, or container near the placards.

Note: When the words INHALATION HAZARD appear on the placards, the INHALATION HAZARD mark is not required on the bulk packaging.

Commodity Name For intermodal tanks transporting any hazardous materials and for tank cars transporting certain hazardous materials, the commodity name must appear on two opposing sides of the intermodal tank or tank car. The commodity name (3.9 in. in height for tank cars and 2 in. in height for intermodal tanks) must match the proper shipping name on the shipping papers and may include the technical name, although it is not specifically required.

Tank Car Specification and Qualification Dates Stencils

- Make sure the stencils describing the tank car specification (e.g., DOT 111A100W1) and qualification dates are legible (see Figure 4.8). These stencils will appear on both sides of the tank car toward the end on the right as you face the car.
- Make sure the tank car qualification dates for pressure (Association of American Railroads).

Hazard Communication/GHS (Figure 4.45) Hazard communication/GHS are mostly used for small containers in the workplace to warn employees the hazards of chemicals they work with on a regular basis. Symbols indicating the hazards will be placed on the label of containers of hazardous materials (United Nations).

Military The military also has a placard system for explosive materials, chemical agents, and biological materials (Figure 4.46). Military placards are mostly fixed-facility markings, with a smaller scale in transportation on military reservations. That system uses orange placards of various shapes to indicate the levels of fire and explosion hazards. If your jurisdiction is near a military facility or has mutual aid with one, you should be familiar with the military marking system (FEMA/NFA).

DOT Placards and Labels (Figure 4.47)

DOT Hazard Classes

Hazard class is one of the most important items of information you can have when dealing with a hazardous material. Not only does the hazard class tell you the hazard of a particular placarded material it will tell you the physical state, which can usually be verified by the type and shape of the container.

Hazard Class 1 – Explosive

Division 1.1 Mass Explosion Hazard
Division 1.2 Projection Hazard
Division 1.3 Fire Hazard
Division 1.4 No Blast Hazard

GHS Pictograms	Physical hazards	GHS Pictograms	Health and Environmental hazards
	Explosive; Self-reactive; Organic peroxide		Skin corrosion; Serious eye damage
	Flammable; Pyrophoric; Self-reactive; Organic peroxide; Self-heating; Emits flammable gases when in contact with water		Acute toxicity (harmful); Skin sensitizer; Irritant (skin and eye); Narcotic effect; Respiratory tract irritant; Hazardous to ozone layer (environment)
	Oxidizer		Respiratory sensitizer; Mutagen; Carcinogen; Reproductive toxicity; Target organ toxicity; Aspiration hazard
	Gas under pressure		Hazardous to aquatic environment
	Corrosive to metals		Acute toxicity (fatal or toxic)

Figure 4.45 Hazard Communication/GHS.

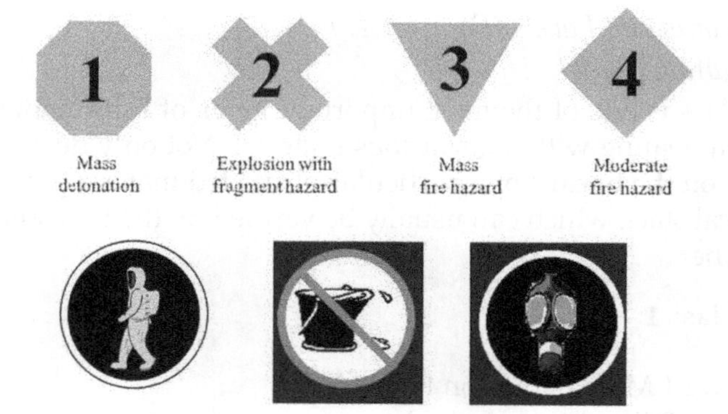

Figure 4.46 The military also has a placard system for explosive materials, chemical agents, and biological materials.

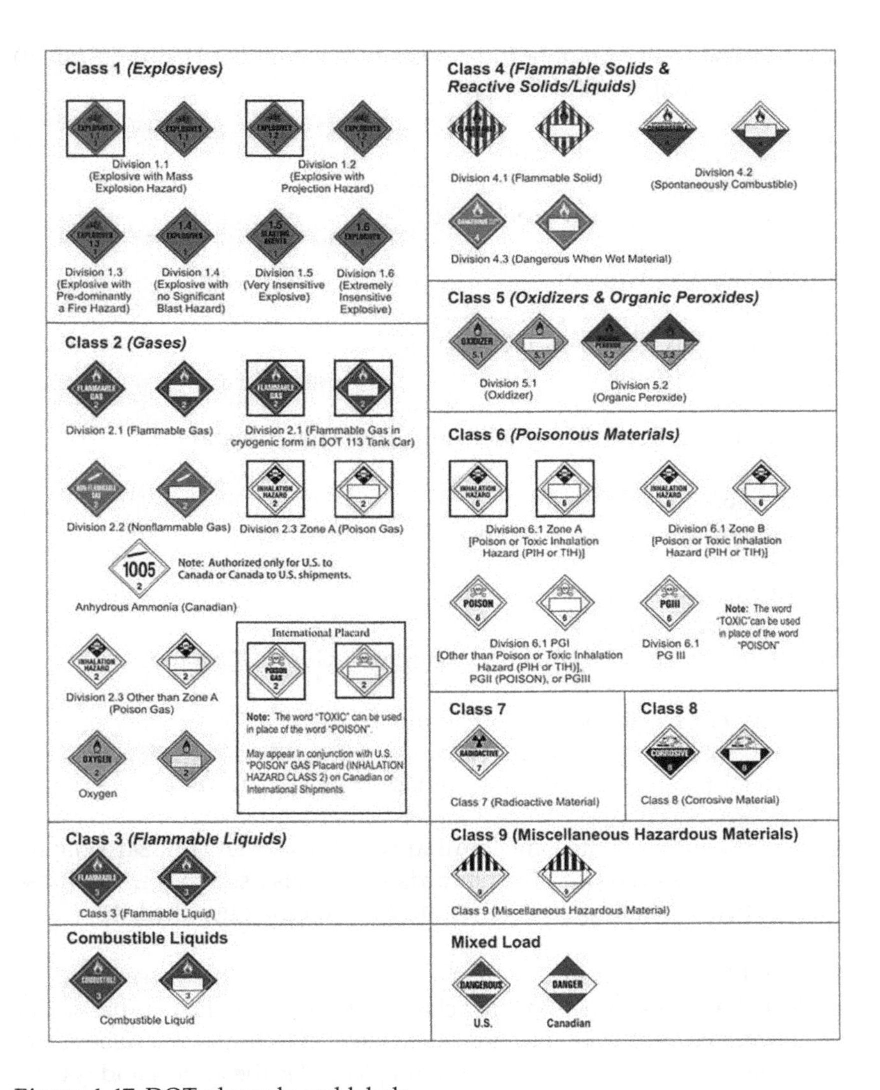

Figure 4.47 DOT placards and labels.

Division 1.5 Blasting Agents
Division 1.6 Insensitive Articles

Hazard Class 2 – Compressed Gases

Division 2.1 Flammable Gases
Division 2.2 Non-Flammable Gases
Division 2.3 Gases Toxic by Inhalation
Division 2.4 Corrosive Gases (Canada)

Hazard Class 3 – Flammable Liquid (and Combustible [United States])
Class 4 – Flammable Solid

Division 4.1 Flammable Solids
Division 4.2 Spontaneously Combustible
Division 4.3 Dangerous When Wet

Hazard Class 5 – Oxidizer

Division 5.1 Oxidizers
Division 5.2 Organic Peroxides

Hazard Class 6 – Toxic Materials and Infectious Substances

Division 6.1 Toxic Materials
Division 6.2 Infectious Substances

Hazard Class 7 – Radioactive
Hazard Class 8 – Corrosive
Hazard Class 9 – Miscellaneous Dangerous Goods (Canada)

Division 9.1 Miscellaneous (Canada)
Division 9.2 Environmentally Hazardous (Canada)
Division 9.3 Dangerous Wastes (Canada)

First responders should become familiar with the U.S. DOT hazard classes for hazardous materials and the placards and labels used to identify those hazards. This information will assist them when they use the ERG.

Placards and Labels

The DOT identifies nine hazard classes. Each class has a particular color associated with its placard and label. Placards and labels are diamond-shaped. A 10 in. × 10 in. diamond placard is the largest and is used on the exterior of transportation vehicles and certain bulk containers. Labels are a 4 in. × 4 in. diamond and found on the individual packages and small containers of hazardous materials. Also located on each placard and label is the hazard class number. This number is located on the bottom corner of the diamond. A symbol is found in the top corner of the diamond, and on most of the placards and labels a hazard class name is located in the center.

DOT regulations permit the use of wordless placards, which are used throughout the rest of the world. Wordless placards do not have the hazard class name in the center of the placard, so responders should become familiar with the colors of placards and labels and the hazard class associated with each color. Many times, the nature of hazmat incidents prevent emergency personnel from getting close enough to read what is written on the placards. However, the color of the placard can be identified from a safe distance or through binoculars. By associating the color of the placard or label with the hazard class, a responder can identify the hazard even if nothing is written on the placard.

Hazardous materials transported by rail, highways, and waterways may be found with placards or labels. Usually, air shipments present a limited hazard, are in small packages, and are labeled. Pipelines are also a type of transportation system for shipping hazardous materials. Pipelines aren't placarded or labeled, but locations are marked with aboveground signs that identify the product and the pipeline company and provide emergency contact information (Figure 4.48).

Hazard Class 1 Explosive The DOT identifies six divisions of explosives with orange placards and labels. While these divisions are identified by certain explosive characteristics, it is important that response personnel do not relate any of the classes with a lesser hazard than another. Responders may not know if the circumstances are present for an explosive to behave in a certain way. Therefore, all explosive classes should be treated as the worst case, that being detonation, until explosive experts arrive on the scene.

Hazard Class 2 Compressed Gas Compressed gas hazard class includes three divisions, flammable, nonflammable, and poison. Compressed gases are a hazard class because of the pressure in the containers, which presents a hazard in addition to the physical and chemical characteristics

Figure 4.48 Pipelines aren't placarded or labeled, but locations are marked with aboveground signs that identify the product and the pipeline company and provide the emergency contact information.

of the gases. Red placards indicate flammable. Flammable gases include propane, hydrogen, and butane. Nonflammable gases have a green placard indicating a compressed gas that is considered by DOT definition to be nonflammable.

> ***Hazmatology Point:*** *This can be dangerously deceiving because anhydrous ammonia is placarded as a nonflammable compressed gas when, in fact, it will burn under certain conditions, usually inside a building or in a confined space. This is because the DOT definition of a flammable gas does not fit the flammable range of anhydrous ammonia.*

Because oxygen is a nonflammable gas and an oxidizer, it may have a yellow placard and label because it is an oxidizer. Cryogenic liquids that are inert do not fit into any other hazard class, and for the purposes of Hazmatology do not react chemically with any other materials. So they are considered nonflammable compressed gases.

> ***Hazmatology Point:*** *Cryogenic liquids do have physical hazards that can pose a dangerous risk to emergency responders. They are extremely cold, have wide expansion ratios, and like all liquefied gases, can cause simple asphyxiation.*

Other nonflammable gases include carbon dioxide, nitrogen, and argon. There is also a white compressed gas placard for poisons. Poison gases include chlorine and phosgene.

Hazard Class 3 Flammable Liquid Flammable liquid placards and labels are red. Flammable liquids include gasoline, acetone, alcohol, and ketone. When response personnel recognize that a flammable hazard exists, they should take every precaution to prevent ignition of the material. Ignition sources include open flames, smoking, welding and other hot operations, heat from friction, radiant heat, static electrical charges, electrical sources, mechanical sparks, and spontaneous ignition. Fire apparatus can be a source of ignition and should be positioned properly.

Hazard Class 4 Flammable Solid Flammable solid materials are Hazard Class 4 and have three different divisions and different placards based on general hazards. The first is flammable solid, with a red and white striped placard. An example is a highway flare. Next are flammable solids that are spontaneously combustible with a white over red placard. Even though the hazard class is flammable solid, spontaneously combustible liquids are also included because they don't fit anywhere else in the hazard class system. One common spontaneously combustible material is phosphorus.

The third flammable solid placard is entirely blue and indicates a material that is "dangerous when wet" and reacts with water, releasing gases, producing splattering, heat, and in some cases, violent explosions.

The heat released can be sufficient to ignite flammable gases released such as hydrogen. These water-reactive materials include sodium and potassium.

Class 5 Oxidizer There are two divisions of oxidizers, 5.1 and 5.2. Division 5.1 oxidizers have yellow placards and labels. Oxidizers include swimming pool chemicals and ammonium nitrate. Division 5.2 oxidizers have a placard with the upper half of the diamond and the lower half of the diamond yellow. They are organic peroxides which are reactive and can undergo polymerization. Organic peroxide materials have a Self-Accelerating Decomposition Temperature or SADT. Refrigeration keeps them from reaching their SADT. Example isobutylene.

Class 6 Poison Class 6 materials are liquid and solid poisons with white placards. Class 6 is subdivided into 6.1 and 6.2 materials. Division 6.1 are poisons, including liquids that are very volatile, but not gases, and military and terrorist nerve and mustard agents. Division 6.2 are biological materials, infectious substances and organisms capable of causing diseases. Example include Ebola, HIV, and hepatitis.

Class 7 Radioactive Radioactive placards are Class 7 and are two tone yellow over white. Three radioactive labels are placed on packages according to the amount of radiation emitted by the materials.

Class 8 Corrosive Class 8 materials are corrosive with a white over black placard. Corrosive materials can be acids or bases and include sulfuric acid and potassium hydroxide. The reason two totally different chemicals that are violently reactive when they come together can be in the same hazard class is if they both are corrosive and will cause the same injuries upon human contact and when they contact other materials.

Class 9 Miscellaneous hazardous material Class 9 materials are miscellaneous hazardous materials with black and white stripes over white. Miscellaneous materials include sulfur and hazardous waste, lithium batteries, and others.

DOT regulations, including the hazard classes, placards and labels, and others, undergo periodic review and revision. Response personnel need to watch for changes in transportation regulations. Fixed facilities are also required to maintain the DOT placard and labeling system for storage and use of hazardous materials. OSHA requires that all placards and labels during transportation remain in place during storage and use. The markings can be removed legally only after a container has been emptied and purged of the hazardous materials and has been disposed of properly (DOT).

Shipping Papers Shipping papers can be found with four of the five DOT transportation systems.

Railroad Shipments With railroad shipments, the shipping papers are called the **Way Bill or Consist** and are located in the train engine (Figure 4.49), usually under the control of the conductor. Most trains no longer have a caboose for the conductor to ride in.

Highway Shipments Highway shipments have shipping papers called the **Bill of Lading or Freight Bill** and are located in the cab of the truck, usually in a slot in the driver-side door (Figure 4.50).

Figure 4.49 With railroad shipments, the shipping papers are called the Way Bill or Consist and are located in the train engine.

Figure 4.50 Highway shipments have shipping papers called the Bill of Lading or Freight Bill and are located in the cab of the truck, usually in a slot in the driver-side door.

Water Shipments Water transportation vessels have shipping papers called the **Dangerous Cargo Manifest**, which is located in the wheelhouse. If a barge is involved, the shipping papers are also found in a small container on the barge (Figure 4.51). Barges may also have a red pennant flying indicating that it is carrying hazardous materials. Even though the pennant is red, the color doesn't identify a particular hazard class.

Air Freight Aircraft transporting hazardous materials have shipping papers called the **AirBill**, which are located in the cockpit of the aircraft with the pilot (Figure 4.52) (FEMA/NFA).

A format of a shipping paper can be found inside the front cover of the ERG, with an example of the type of information it contains. Typical shipping papers provide an emergency contact number for the manufacturer of the hazardous material, number and type of packages, chemical name, hazard class, quantity of product, and 4-digit identification number. Quantities will be listed using the metric system of weight, so response personnel should have a working knowledge of the metric system. The shipping papers may also identify a quantity of material, which must be reported to the National Response Center (NRC) if released, and will be marked with an "RQ," indicating a reportable quantity. Materials with an "RQ" are on the U.S. EPA list of materials that must be reported if released into the environment. The emergency contact number at the top of the first page of a shipping paper should be called for response information rather than CHEMTREC. If the shipping papers are not available, then CHEMTREC should be called.

MSDS/SOGs are found at fixed facilities storing or using hazardous materials. Certain facilities with extremely hazardous materials identified

Figure 4.51 Water transportation vessels have shipping papers called the Dangerous Cargo Manifest, which is located in the wheelhouse. If a barge is involved, the shipping papers are also found in a small container on the barge.

Figure 4.52 Aircraft transporting hazardous materials have shipping papers called the AirBill and are located in the cockpit of the aircraft with the pilot.

by the U.S. EPA and Local Emergency Planning Committee (LEPC) are required to submit MSDS sheets to the local fire department, and information about the facility is available from the LEPC. The LEPC was created as a result of the Emergency Planning and Community Right to Know Act of 1986, also known as EPCRA. This act allows local fire departments to access any facility in the community which stores or uses hazardous materials for the purpose of preplanning a hazardous materials release. MSDS contain chemical names, synonyms, hazards, protective equipment, spill cleanup information, and decontamination information.

CHEMTREC has the largest database in the world with 2.8 million MSDS, which is another good reason to contact them right away when you arrive on a hazardous materials scene.

Human Senses

Sight Looking at hazardous materials, especially through binoculars or through the lens of a robot or drone, is one of the safest methods of reconnaissance for sizing up an incident scene.

Hearing Listening for sounds that might be present from a safe distance is also a good method. Witnesses of propane tanks on fire have told stories about the roaring jet engine sound of the relief valve on a tank. When you hear that sound it is your key to evacuate as far away as the DOT ERG tells you to. Then, take video and photographs of the BLEVE. Other sounds can also give you an idea of what is happening at an incident scene, such as a flowing liquid, dripping leak, or hissing of a nonpressure container that has come under pressure from an actor at a hazardous materials scene.

Sight and hearing are the only safe methods of surveying a hazardous materials scene, as long as they are done from a safe distance. The next three senses are just plain dangerous to use and should not be considered for any reason.

Smell

> ***Hazmatology Point:*** *Did you ever notice that some reference materials tell you what a chemical smells like? Doesn't that make you wonder who found that out and if they are still alive? Using smell, unless a witness inadvertently smelled something, is dangerous to responders and unsuspecting witnesses. Keep your nose behind the protection of your SCBA or a safe distance from harm's way. There is a Firefighter Safety Video floating around that was created by the National Fire Academy. It shows a battalion chief on a big city fire department walking up to a leaking tank car in a rail yard and touching the liquid with his hand and smelling the liquid.*
>
> *I always wondered who took the video. Well, one day I found out that Geraldo Rivera, an investigative reporter, just happened to be filming a documentary that day in the rail yard about train accidents and thankfully filmed and shared it for us to learn from.*

Taste

> ***Hazmatology Point:*** *The same goes for taste. Reference sources also have a habit of reporting how hazardous materials taste. Don't want to know who did the tasting or if they are still around. Responders should certainly not be tasting those materials to see if they have a match.*

Touch

> ***Hazmatology Point:*** *Touch is another no, no when dealing with hazardous materials unless responders are wearing a glove that protects the responder from injury from a particular material. Even if you have the proper protective equipment, you don't need to be coming in contact with hazardous materials unless it is absolutely necessary to achieve incident mitigation goals. If you do not touch or come in contact, common sense would tell you that you are not contaminated and you do not require decontamination.*

DOT Emergency Response Guidebook

The first 15–20 minutes of a hazmat transportation accident are the most critical. According to the DOT it is an invaluable resource during the initial phase of a transportation incident involving dangerous goods/hazardous materials. (Figure 4.53). However, the DOT ERG also has helpful information that can be used through out an incident, such as the evacuation

Figure 4.53 The first 15–20 min of a hazmat transportation accident are the most critical.

and isolation distances and charts. Also contact information located in the guide. The U.S. DOT, the creator and distributor of the ERG, celebrated its 50th year since it was established during 2016. More than 1.5 million free copies have been provided to first responders nationwide by the DOT through state contacts in each state. Earlier this year I spoke with the DOT Pipeline and Hazardous Materials Safety Administration (PHMSA) about the latest edition of the ERG. My main focus was "What in particular would they want to have first responders know about the ERG?" Their response was, "emphasize the fact that the ERG is intended for use by first responders during the initial phase (first 30 min) of a transportation incident involving hazardous materials and is not intended for use during incidents at fixed facilities." Additionally, DOT also wanted responders to **RESIST RUSHING IN!** Hazardous materials incidents do not occur often in many of the jurisdictions throughout the United States, Canada, and Mexico. They are technical in nature and can be extremely dangerous to response personnel. It is important that first responders exercise competencies outlined by OSHA 1910.120 and NFPA Standards 472–473, which are designed to keep them safe at the scene of a hazardous materials incident while dictating their limitations in terms of training and equipment.

Each issue of the ERG is on a 4-year cycle and work begins on the next edition soon after the current one is published. According to the 2020 ERG, working group consists of government representatives from the U.S. DOT/PHMSA; Transport Canada; Secretariat of Transport and Communications in Mexico; Argentina's emergency response call center; CEQUIME, Brazil's PRO-QUIMICA; Colombia's CISPROQUIM and Chile's CITUC QUIMICO. Comments are solicited from ERG users and stakeholders through an announcement in the Federal Register.

Changes and Improvements A number of changes are made to the ERG every 4 years to make it more useful to emergency response personnel. On the cover is the statement that sets the tone for the use of the ERG: "A guidebook intended for use by first responders during the initial phase of a transportation incident involving dangerous goods/hazardous materials."

Evolution of the Emergency Response Guidebook
The field of hazardous materials is a relatively new subject for emergency responders. DOT developed the first version of a responder guidebook in 1977 (Figure 4.54). It was called the Hazardous Materials Action Guides. It was 87 pages long and contained information on 43 chemicals. During 1980 the DOT created the first version of the ERG utilizing the format we are familiar with today. It contained 66 orange guides, placard chart, numerical, and alphabetical sections, although they were not yet color-coded as they are in later versions. There have been thirteen editions of the ERG to date: 1980, 1984, 1987, 1990, 1993, 1996, 2000, 2004, 2008, 2012, 2016 and 2020. By comparison, the 2020 version of the ERG contains 392 pages and lists hundreds of chemicals. During 2024 the next version of the ERG will be published. The quantity of information available to first responders for dealing with the initial stages of a hazardous materials incident compared to the early books is enormous. It is important that response

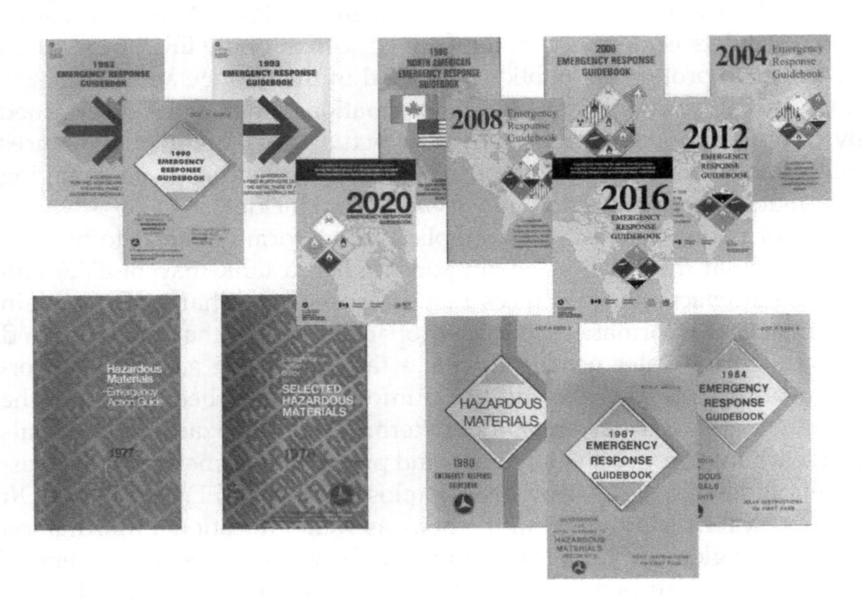

Figure 4.54 The field of hazardous materials is a relatively new subject for emergency responders. DOT developed the first version of a responder guidebook in 1977.

personnel be trained on the use of the book and be very familiar with the book in order for it to be of greatest benefit at an incident.

Initial Response Actions

The ERG addresses all first responder duties and provides resources to help carry them out. The ERG has many resources to help in recognition, including the Placard Charts, Rail Car Charts, Road Trailer Charts, GHS Labels, and Pipeline section. Once it is determined that a hazardous materials incident has occurred, in general, first responders do not have the necessary training or equipment to mitigate the incident. Response personnel need to access the Yellow and Blue sections of the ERG to determine a name for a hazardous material or find a generic Orange Guide Page based upon placards, Tank Car, Road Trailer, or pipeline information in the ERG. Within the ERG are several notification resources including CHEMTREC, the NRC, military shipments, National Poison Control Center, as well as some private response companies. Also, in the ERG is a page you can fill in with local response contact information, including local railroad contacts. Following proper notifications, the incident site needs to be isolated so that responders and the public do not come in contact with the hazardous material. ERG contains information on initial isolation distances in the Orange and Green Guide sections.

First responders need to protect themselves and the public from harm created by the release of a hazardous material. Protection information for responders is identified in the Orange section of the ERG. Evacuation distances to protect the public are located in the Orange and Green sections. Initial isolation distances and evacuation distances are determined by computer modeling and analysis of actual incidents; some distances may change from one edition of the ERG to another based on changing information from computer models and actual incidents.

Information has also been supplied in the form of charts to provide supplemental information on situations when a tank may BLEVE, and additional evacuation distances may be necessary. Charts also contain some tactical information such as propane behavior characteristics and the amount of water needed to cool a tank. With the advent of terrorism over the past 25 years, additional information has been added to the guidebook to assist responders in determining if a hazardous materials incident could be an act of terrorism and provides information on terrorist agents and devices. An Improvised Explosive Device (IED) Safe Stand-Off Distance Chart is provided in the book as well. Educational information including a glossary has been provided in the white pages near the end of the ERG to help responders understand how the book is organized

The ERG is a valuable tool to identify hazardous materials and determine actions to be taken by first responders in the first 30 min of a dangerous goods/hazardous materials incident. While the DOT stresses that the

ERG is just for transportation incidents only, there is information in the ERG that can be useful for any hazardous materials incident, particularly if the ERG is the only information resource available during the initial phases of an incident. Just as first responders are limited in what they can do at the scene of a hazmat emergency by equipment and training, the ERG is limited in the amount of information it provides. ERG was not intended to be used during the mitigation phase of the incident, and therefore, should not be used as one of the reference materials selected to determine mitigation methods.

Free copies of the ERG are provided by the DOT to all fire, police, EMS, and other emergency response organizations through a selected agency in each state. Your state agency can be determined by contacting the DOT Office of Hazardous Materials Transportation Research and Special Programs (RSPA) at 202-366-0656. State contacts are also listed on the DOT ERG section of the website. PSMSA partnered with the National Library of Medicine (NLM) to provide a free electronic version of the ERG for emergency responders. Copies are also available for a fee from private companies online. The author has developed a training course for the ERG that is available in PowerPoint CD-ROM. Also included are an instructor guide and student manual, list of state ERG contacts, list of private-sector sources for the response guide, a course certificate template, a final exam, and an electronic version of the ERG (DOT). For information about the "Placard Hazard Chart" or the ERG training course, contact Robert Burke at robert.burke@windstream.net or visit www.hazardousmaterialspage.com.

Notification

Notification is the process of identifying additional resources needed at an incident and making the call. Notifying others is required because first responders do not have the training or equipment to mitigate the circumstances surrounding a hazmat or terrorist incident. Responders need to know who in their jurisdiction and who outside their jurisdiction needs to be notified when an incident occurs. The dispatcher may have the contact list and protocols to make the contacts. Even if the only call they personally make is to the dispatcher, responders still need to be aware of the resources that will be needed to respond to the scene.

National Response Center

The U.S. Coast Guard (USCG) operates the NRC, which is a part of the Department of Homeland Security during peacetime. This is the contact point for federal assistance for terrorist and hazmat incidents and can be reached 24 hours a day at 800-424-8802.

If response agencies want to seek reimbursement from the U.S. EPA for expenses incurred at hazmat incidents, NRC must be contacted.

EPA has a program to reimburse a single agency up to $25,000 for eligible expenses incurred during an incident that cannot be recovered from the spiller or another source. If NRC is not called soon after the incident occurs, reimbursement cannot take place. In addition to federal contact information, NRC has a chemical database and teleconferencing capability to assist responders at the incident scene (FEMA/NFA).

CHEMTREC

Do You Have an Emergency Involving Chemicals? CHEMTREC is a 24-h emergency center that was started as a service of the Chemical Manufacturers Association (CMA), now known as the American Chemistry Council. The first center was located in downtown Washington, DC. It is now located across the Potomac River in Arlington, VA. In addition to its chemical emergency call center, CHEMTREC provides training programs, MEDTREC (for chemical medical information), MSDS and CHEMNET, an industry chemical emergency mutual aid system.

Emergency Call Center

During an interview with the Director of the Operations Center, Joe Milazzo, it was interesting to find that the call volume to the emergency call center is down. They currently handle 99,000 cases and 55,000 actual emergency calls per year, an average of 350 calls per day. He believes the reason is easy access to online information. It is important to note that CHEMTREC should still be your first call before searching electronic resources (Figure 4.55). There are a number of advantages to calling them first. One call can put responders in touch with the company that

Figure 4.55 It is important to note that CHEMTREC should still be your first call before searching for electronic resources.

manufactured a chemical, the shipper, provide MSDS, and activate and send a chemical company or industry mutual aid hazmat response team to the scene. CHEMTREC can also provide on-scene responders with information on over 1 million chemicals. Procedures are in place for calls originating from non-English-speaking callers. Specialized assistance from chemists can be obtained from its extensive database of chemical industry contacts. Emergency collect calls are accepted, and all calls are recorded.

CHEMTREC provides emergency chemical information to emergency responders nationwide. Its primary mission "is to provide technical information about products involved, guidance on how to protect themselves and the public, and what initial action is required to mitigate the incident." CHEMTREC strives to provide this and other information quickly and accurately 24h a day. Emergency response organizations do not have to be registered with CHEMTREC to use its services. Information and assistance are provided free of charge as a public service to emergency response organizations. Backup power is provided for the center and protocols are in place for relocating the center in an emergency.

CHEMTREC History

CHEMTREC began operations on September 5, 1971. In the beginning, CHEMTREC Communicators, as they were called, were merely conveyors of information (Figure 4.56). Many were former military personnel who were recruited because of their ability to keep cool during crises. Chemical data was kept in a sophisticated card file system, and the Communicator would read the information on the card with little, if any, additional input (Figure 4.57). They were not required to have any expertise in the areas of hazardous materials or emergency response.

Figure 4.56 CHEMTREC began operations on September 5, 1971.

Figure 4.57 Chemical data was kept in a sophisticated card file system and the Communicator would read the information on the card with little, if any, additional input.

- 1918

  Responding to a series of railway accidents involving shipments of corrosive liquids vital to war efforts, the Manufacturing Chemists' Association (MCA) forms a committee devoted to the improvement of liquid chemical shipping containers.
- 1932

  MCA begins a comprehensive program of Safety Manuals and Safety Data Sheets. Each datasheet details a product's physical and chemical properties, its hazards, instructions on safe handling and first-aid required labeling or identification, and methods of unloading and emptying various types of containers.
- 1969

  The U.S. DOT meets with MCA to determine the best approach for reporting and responding to emergency situations involving chemicals in transport.
- 1970

  MCA establishes CHEMTREC to provide chemical-specific information to emergency responders around the clock. Members authored over 1,600 chemical cards representing the chemicals most frequently involved in transportation.
- 1971

  On September 5th, CHEMTREC becomes fully operational.
- 1978

  MCA becomes the Chemical Manufacturer's Association (CMA).

- 2000
  CMA becomes the American Chemistry Council (ACC).
- 2004
  CHEMTREC launches medical service coverage.

The current version of the Communicator is the Emergency Services Specialist (ESS), highly trained and qualified in hazardous materials and emergency response. ESS personnel come from the emergency services with experience in emergency response, firefighting, EMS, military emergencies, and related fields. The old card file system has been replaced with a computerized database and ESSs provide additional information and recommendations based on their knowledge and experience dealing with hazmat emergencies.

CHEMTREC can provide technical information directly to the scene of a hazmat incident, including faxed information and MSDS from its document library. It also provides information to emergency medical personnel concerning chemical exposure treatment. ESS personnel gather information about the incident, product, and exposure, and then provide medical treatment information from the MSDS (Figure 4.58).

You can call CHEMTREC right after you arrive on scene and they will be able to assist you. However, for CHEMTREC to provide information to the response scene in a timely manner, it is suggested that responders gather the available information needed by CHEMTREC before making a call to them. The following is some of the suggested information.

Figure 4.58 Captain Francisco Martinez from the Lincoln, NE Hazmat Team looking over an MSDS sheet in preparation for determining action options. (Courtesy: Lincoln, NE Fire Department.)

- Caller's name and rank or title.
- Caller's company or organization.
- Caller's location.
- At least one callback number, with the area code.
- Dispatch center number, if available.
- Fax number or email address.
- Location of incident/weather conditions.
- Time the incident occurred.
- Type or description of container/package.
- Container numbers and/or markings.
- Brief description of the incident and the actions taken.
- Number and type of injuries/exposures.
- Amount of product(s) involved and released.
- Is there specific information needed as a priority?
- Are any industry representatives on the scene or have any been contacted?

If shipping papers have been obtained, the following additional information is requested by CHEMTREC:

a. UN/NA Identification Number (Placard) or STCC number of the products.
b. Chemical name, product(s) name, or a trade name.
c. Carrier name.
d. Shipper and point of origin.
e. Consignee and destination.

Not all of the incidents received by CHEMTREC are massive train derailments or tank truck accidents. As many as 50% of the incidents called into the emergency center involve 5 gallons or less of product. Below are examples of two typical and one unusual incident handled by the call center. (Sample CHEMTREC incidents are from the User's Guide for Emergency Responders.)

Incident 1 – CHEMTREC received a call from a railroad reporting a train derailment. The report stated that six cars containing sulfuric acid and two empty cars that had carried naphtha/xylene mixture and silicon tetrachloride were involved. An unknown amount of product was reported to be leaking from the rail cars. Adverse weather conditions in the area were hampering response and evaluation of the incident. CHEMTREC faxed MSDS, and then discussed the situation with the local fire department and the nearest poison control center. CHEMTREC also conferenced with the shippers as well as the fire department and railroad representatives.

Incident 2 – A trucking company dispatcher called in to report that an unknown product was leaking out from the trailer at a truck stop in Arizona. The local fire department was notified who contacted CHEMTREC for assistance identifying the products in the trailer. As there were multiple shippers' products on board, CHEMTREC asked that the bill of lading be faxed to the emergency call center. The shipping papers received indicated that two shippers were involved, shipping hydrofluoric acid solutions and ethylene glycol. The ESS contacted each shipper, requesting contact with the reporting carrier's dispatcher. In turn, shippers called back to advise that their respective emergency coordinators had made contact and that product information had been supplied to the responders.

Incident 3 – While the author was visiting CHEMTREC to obtain information for this column, a call was received from a state National Guard and logged by an ESS. An auctioneer found a container marked as chemical weapon gas. The container was also marked "Manufactured by Lake Erie Chemical." A bomb squad had already entered the building and was able to view the container. The container was described as a plastic cylindrical object approximately 15 in. in length. There were two silver vials inside with a percussion cap. There was a brown opaque liquid in the vials. Cotton padding was in place between the two vials. The container was marked as "Lot 2250."

The ESS researched Internet websites to obtain the following information, which was sent to the National Guard. During the 1930s, the Lake Erie Chemical Company, based in Cleveland, developed a system to discharge tear gas into bank lobbies during holdups. This was done in conjunction with a company called Diebold, which was run by Elliot Ness, of "The Untouchables" fame. Lake Erie Chemical also manufactured flare pistols. One of the items in the tube was described as having a percussion cap on its base. CHEMTREC advised the National Guard of the strong possibility that the item was a dispensing unit that may contain tear gas. The item was collected and secured for disposal by the local bomb squad.

Assistance for Responders

Shipping Paper Emergency Contact Information

The U.S. DOT requires that a shipper of hazardous materials provide a 24-h emergency contact number at the top of all shipping papers. If a company does not have the capability of providing this type of contact, CHEMTREC can contract with the company and provide the emergency contact service. Therefore, some shipping papers will have the CHEMTREC 24-h emergency number at the top for emergency contact information. Others will have a number directly connected to the shipper or manufacturer. Response personnel should use the shipper or manufacturer number first, if provided, before contacting CHEMTREC.

MSDS system – CHEMTREC has one of the largest databases of MSDS in the world. Over 2.8 million documents are available and can be faxed to emergency response personnel on scene of a hazmat incident. The MSDS system at CHEMTREC was recently upgraded to allow for better management and quicker retrieval of documents during an emergency. Information can be faxed directly to the incident scene. Response personnel should identify the location of the nearest fax machine and provide the number to CHEMTREC to receive information.

Participation in drills and exercises – Through prior arrangement, CHEMTREC participates in local hazmat drills and exercises. Using the emergency call center during a drill or exercise can help emergency responders to better understand the resources and services available during an actual emergency. Participation by product shippers or manufacturers can also be arranged for specific chemicals.

MSDS for materials involved in a drill or exercise can be obtained prior to the event by contacting CHEMTREC's nonemergency number. Request forms for CHEMTREC participation are available by calling 800-262-8200 or can be downloaded from the website. Completed forms should be faxed or e-mailed at least 48 h prior to the drill or 10 days prior for regular mail. Mail should be sent to CHEMTREC emergency call center, 1300 Wilson Blvd., Arlington, VA 22209-2380. Call 703-741-5525 at least 1 day in advance of the drill to confirm that the registration was received (CHEMTREC).

CHEMTREC Website

A section of the website is dedicated to emergency responders. This site was improved and updated recently to better meet the needs of users. Current information is provided on training opportunities, regulation changes, reports, and other articles of interest for emergency responders. Helpful links to other hazmat sites are provided along with frequently asked questions.

Protection

Responders should avoid contact with any hazardous material irrespective of its form. Usually, liquid and solid hazardous materials are evident when spilled. Personnel should avoid contact with these materials even if they are wearing proper protective equipment unless absolutely necessary for mitigation of the incident. EMS personnel, police officers, and other responders may not have respiratory or other protection available. These personnel should stay a safe distance from the potentially hazardous material or terrorist agent until personnel in proper protective clothing, using instrumentation, can identify the dangers.

Protection includes both the emergency responders and the public who may be in harm's way of the hazardous material(s). Protection is accomplished by four methods:

- Deny entry to the hot zone.
- Wear proper protective clothing for the hazard present.
- Evacuation.
- Sheltering in place.

The hot zone, or exclusionary zone as it is sometimes called, is where the hazardous material or anything that has contacted the hazardous material is located, including victims. Protection of responders and the public starts with denying entry to the hot zone. Denying entry is simply keeping everyone, including responders, out of the hot zone by placing barrier tape, posting personnel or police, or using public address equipment.

A hot zone should be identified, and entry should denied to everyone on scene, including responders who are not properly trained or equipped to enter. Victims need to be decontaminated before EMS personnel provide treatment. First responders at the operations level can perform emergency decontamination to remove hazardous materials to reduce the impact on the victims. Victims, however, should still go through technical decontamination before treatment by EMS personnel. Hazardous materials can cause injury or death to responders and the public in a number of ways. Hazards in the hot zone include:

- Thermal, hot or cold
- Mechanical
- Toxic
- Corrosive
- Asphyxiation, both chemical and simple
- Radiation
- Etiological (biological) and infectious substances

Hazardous materials and terrorist agents can enter the human body through four routes:

- Inhalation
- Ingestion
- Absorption
- Injection

Wearing proper protective clothing or maintaining a safe distance will help reduce the chance of these types of exposure (Figure 4.59). Incident scenes should be approached from upwind, uphill, and upstream, whenever possible, but always upwind. Responders should make use of protective equipment available to them or withdraw from the area if they have no protection. Respiratory protection is the most important concern when dealing with potentially hazardous materials or terrorist incidents.

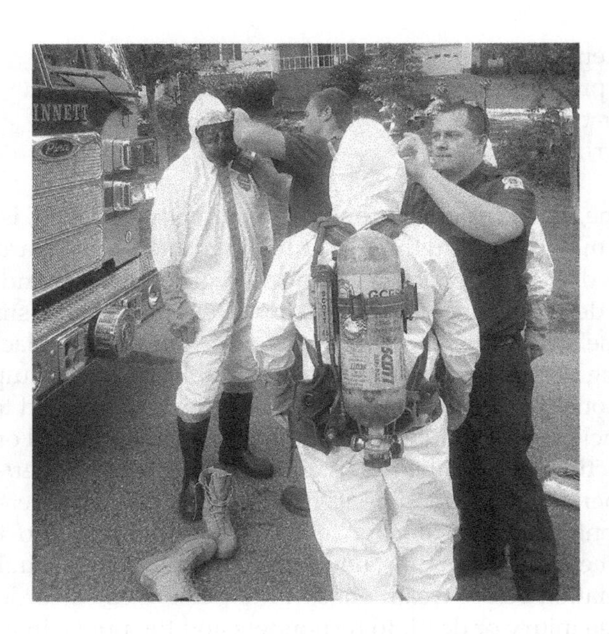

Figure 4.59 Wearing proper protective clothing or maintaining a safe distance will help reduce the chance of exposure to hazardous materials. (Courtesy: Gwinnett County GA Fire Department.)

All responders, including fire, EMS, special operations, and hazmat, should always have access to SCBA.

If there is any chance by identification, location, or other information that hazardous materials might be present, **ALWAYS** wear SCBA until the atmosphere has been proven safe by monitoring. Materials such as gases and volatile liquids, which present respiratory dangers, are the most serious of the hazardous materials that are out there. Not all airborne hazardous materials are visible. Firefighters generally have the highest level of respiratory protection with their SCBA. This protection should be worn in conjunction with full firefighter turnouts when the potential for hazardous materials or terrorist agents is determined. EMS personnel should also have access to SCBA on their own apparatus.

Persons exposed to hazardous materials can also experience psychological harm or believe they have been exposed to the materials and even experience symptoms, when in fact they have not contacted the hazardous material. Emergency decontamination can be undertaken by first responders, because with hazardous materials, the longer they are in contact with the human body the more damage is done. Emergency decontamination will be discussed further in this volume.

If first responding emergency personnel do not follow proper awareness procedures, they may also become victims and may have to remain in the hot zone to await decontamination. Anyone who enters the hot zone has the potential to become contaminated and may need to undergo decontamination before exiting. Equipment and protective clothing may also become contaminated and may need to undergo decontamination or disposal before leaving the hot zone. If this potential contamination occurs before the materials have been identified or it is known for sure no contamination took place first responders should do an emergency decontamination just in case.

Protecting the Public

In addition to protecting themselves, first responders must also protect the public from contact with hazardous materials. Besides establishing the "hot zone" and denying entry to the public, responders may also have to evacuate the public or shelter people in place.

Evacuation and isolation area distance information can be obtained from the green- or orange-bordered pages of the ERG. The green section provides information for materials that pose a health hazard. Inhalation is the primary route of exposure, and this section is used only when no fire is involved. Distances provided in the green section are based on computer models of small and large spills and data from previous incidents. Spills are further classified by daytime and nighttime occurrences.

Because air is more stable at night, vapor clouds can remain together for longer periods and travel greater distances than in the daytime. When materials are on fire, the orange section of the ERG is used to determine isolation and evacuation distances.

Training on the proper use of the ERG is very important for first responders. Resources are limited in the first minutes of a hazmat or terrorist incident. Responders may only be able to assist persons who are immediately in harm's way. Others may have to be moved or sheltered as the incident progresses and more resources are available to assist in protective measures and public notification.

Evacuation

If the public is to be evacuated from homes or businesses, an evacuation center needs to be established to receive and care for the needs of the evacuees. Workplaces and schools may have internal emergency plans that should be considered when evacuation or sheltering in place involves their facility. Emergency management agencies and LEPCs have developed plans for the entire community covering notification, evacuation, and sheltering-in-place procedures to be used during hazmat incidents. Evacuation plans should identify evacuation centers for those who are displaced. If it is determined that a terrorist incident has occurred, published evacuation plans may need to be modified especially concerning evacuation centers as they may become secondary targets.

State and local emergency management agencies also have the resources to notify the public through the Emergency Action System (EAS) (formerly the Emergency Broadcast System), sirens, and radio and TV broadcasts. By contacting the emergency management agency, responders can set in motion procedures for notifying and protecting the public. When evacuations are necessary, preparations must be made to evacuate pets and protect livestock, if possible. Evacuation centers need to be equipped not only for people but also pets that are evacuated.

Whether evacuated or sheltered in place, those in danger need to be notified of the danger and given instructions on what to do. People who would be placed in greater danger by evacuation than if they stayed in place can be sheltered. Residents of communities should be encouraged to have disaster supplies on hand in the event they have to shelter in place. These include:

- Flashlight and extra batteries
- Portable, battery-operated radios and extra batteries
- First-aid kits and manuals
- Emergency food and water
- Nonelectric can openers
- Essential medicines

Procedures should be established for reuniting family members who may become separated during an incident. During the day, parents are at work and children are at school or daycare, so having a reunification plan can be very important. An out-of-state relative should be established as a contact so that all family members know whom to call in the event they become separated (FEMA/NFA).

CASE STUDY

MISSISSAUGA, ON, CANADA NOVEMBER 10, 1979 DERAILMENT FIRES EXPLOSIONS

The "Mother of All Hazmat Evacuation"

Mississauga Train Derailment, Mississauga, ON, 10 November 1979: A 106-car Canadian Pacific Railway train was eastbound near Mississauga, Ontario when a wheel bearing on the 33rd car began to overheat due to a lack of lubrication. Shortly before midnight, a wheel/axle assembly on the car fell off of that car causing 23 cars to derail. The derailed cars contained a variety of cargo including styrene, toluene, propane, sodium hydroxide, and chlorine.

Figure 4.60 The derailed cars contained a variety of cargo including styrene, toluene, propane, sodium hydroxide, and chlorine. (Courtesy: Mississauga Fire Department.)

Several of these cars, including the car carrying chlorine, ruptured and spilled their contents. Propane cars exploded and burned (Figure 4.60). The force of the explosions knocked emergency responders to the ground and hurled one propane car over half a mile. An hour and a half after the derailment occurred officials ordered an evacuation. Additional evacuations over the next 2 days caused more than 218,000 of the 248,000 Mississauga residents to leave the area. Firefighters initially concentrated on cooling cars to allow the fire to burn itself out (Mississauga Fire Department).

BRYAN, TX JULY 30, 2009 AMMONIUM NITRATE FIRE

A fertilizer plant caught fire. Fearing an explosion 80,000 people were evacuated. The fire department did not fight the fire and stayed back at a safe distance. No explosion occurred (Chemical Safety Board).

Shelter in Place

Residents who are asked to shelter in place should be given instructions on what to do to protect themselves. Many communities near major transportation routes or chemical plants encourage residents to put together shelter-in-place kits that contain essential supplies to prepare the home to exclude outside air exchange during an emergency. Shelter-in-place procedures include the following:

- Bring pets and family members inside.
- Close and lock windows and doors.
- Seal gaps under doorways and windows with wet towels and duct tape.
- Seal gaps around windows and air conditioning units, bathroom and kitchen exhaust fans, and stove and dryer vents with duct tape and plastic sheeting, wax paper, or aluminum wrap.
- Close fireplace dampers.
- Close off nonessential rooms such as storage areas, laundry rooms, and extra bedrooms.
- Turn off air handling or ventilation systems.

Sheltering in place is ordered when the atmosphere outside caused by a hazardous materials release is too dangerous for people to be in. Keeping people and pets in their homes will provide a higher level of protection than evacuating them. Tests have been conducted by Utah Valley University at the Dugway Proving Grounds in Utah with releases of chlorine and anhydrous ammonia. Chlorine and ammonia were chosen because of their frequency of atmospheric release and the resulting fatalities. Both chlorine and ammonia are common hazardous materials in communities across the United States and Canada. Data proves that atmospheres inside buildings and vehicles are much less and safer inside than outside (FEMA/NFA).

CASE STUDY

CRETE, NE FEBRUARY 18, 1969 AMMONIA RELEASE

On February 18, 1969, a BNSF train derailed in the center of Crete, releasing 29,200 gallons of liquid ammonia, which almost immediately turned into ammonia gas killing nine people (Figure 4.61). One gallon of ammonia liquid produces 877 gallons of gas volume. It appears that the highest and ultimately lethal concentrations of ammonia were located on the west end of 13th Street and on the south side of the street. All of the victims were in that area when the impact occurred. Three Crete residents died during the accident, three died later in the hospital. Three unidentified transients riding the train were killed by trauma during the derailment. Injury

Figure 4.61 On February 18, 1969, a BNSF train derailed in the center of Crete, releasing 29,200 gallons of liquid ammonia, which almost immediately turned into ammonia gas killing nine people. (Courtesy: Crete News.)

reports varied; however, the Crete News, the local paper, lists approximately 25, although the NTSB reported 53 in its final report. Of the injured were two train crew members. The train conductor fell approximately 18 ft as he stepped from the train, going over a bridge west of the derailment. He was later transferred to Lincoln for treatment.

The idea of sheltering people in place inside of buildings against chemical exposure did not exist at this time. Many people did in fact shelter themselves inside their homes, placed wet rags over their faces, and some covered with blankets. Not only did they shelter in place and protect themselves from ammonia but they saved their lives because of their inadvertent actions. Not a single person who stayed inside the entire time of the emergency and took self-protective actions died or was seriously injured.

Those who died from ammonia exposure died when they left their homes and were overcome by the ammonia vapors outside. Some died on their driveways and one on a street corner, beyond the apparent safety of their homes. Curiosity called them out to see what had happened and they paid the ultimate price. One person who was inside their residence and quickly went onto their porch turned around and went back inside. The person still had enough exposure to the ammonia to die of complications later in the hospital. Unknowingly, even before the concept of shelter in place had been developed, victims of the Crete derailment confirmed the effectiveness of sheltering in place when hazardous materials are released outside of buildings or a vapor cloud travels to populated areas preventing an expeditious and safe evacuation (*Firehouse Magazine*).

Awareness- and Operations-Level Training

To carry out their required tasks, that is, recognition, notification, identification, and protection, awareness-level personnel must be trained, response plans developed, and standard operating procedures (SOPs) written to govern responder actions at hazmat or terrorist incidents (Figure 4.62). As mentioned previously, the OSHA does not define how many hours of training should be provided to response personnel at the awareness level or which training is acceptable for certification. The employer is responsible for identifying training that meets the awareness requirement for competencies. Operations personnel need to be trained to the appropriate competencies with 8h of initial training and 8h of refresher training annually. Employers should then certify the employee to the awareness or operations level as appropriate to their duties on the job. Refresher training must be provided each year.

Figure 4.62 To carry out their required tasks response plans developed, and SOPs written to govern responder actions at hazmat or terrorist incidents.

Technician Level Response

Hazardous Materials Team Tool Box

When a hazardous materials incident occurs and it is necessary to call a hazardous materials team into action, they respond with tools to mitigate the emergency. Tools include physical, cognitive and experience necessary to do the job. Our hazardous materials toolboxes contain many tools used for different situations that may face us upon arrival on an incident scene. Hazmat Tool Boxes may contain different tools in various parts of the country based on the known hazardous materials exposures faced in our jurisdictions. However, like any toolbox, not all tools are needed every time we respond. Just because we have certain tools does not mean we have to use them solely because they are available.

For example, decontamination is one of our tools (Figure 4.63); but it does not mean we need to perform decontamination every time we respond just because we can or for having an overabundance of caution. Common sense needs to be exercised when deciding if decontamination is needed. Decontamination is performed because someone or something has become contaminated. Contamination occurs depending upon the circumstances of the incident. It is not difficult to determine if contamination has occurred. In fact much of that determination is science and plain common sense. The frequency and type of decontamination are based heavily on the physical state of the contaminant. Part of our standard operating procedures (SOPs) should dictate that we practice contamination avoidance at all times. Just because we may have personal protective

Figure 4.63 Decontamination is of one of the tools in our Hazmat Tool Box.

equipment (PPE) to protect us from a hazardous material doesn't mean we have to intentionally contact that material.

PPE is another tool in our Hazmat Tool Box (Figure 4.64). We carry several types of PPE that are selected for use based on the hazards of the incident scene. Generally, the hazards we expect to encounter are based on a hazardous material that has been released from its container, its physical state and the effects those hazards may have on responders as they go about the mitigation of an incident. Our PPE is designed to protect us from specific hazards. Wearing a higher level of PPE than is required for those hazards **does not** protect us any better than the level that is required. In fact, higher levels of PPE have greater impact on the health of emergency responders, and therefore, should be used **only** when that level of protection is required.

For example, some hazmat teams only wear Level A PPE because that is the highest level of protection; in fact, they use it for everything. There are several flaws in that thought process. First, only around 3% of all hazardous materials incidents require Level A protection because of the hazards of a particular material. So, wearing Level A protection on all incidents does not provide us with any better protection than the appropriate level of PPE for the hazards of a material we are trying to mitigate. Level A PPE is expensive to purchase, replace if damaged through overuse, and labor-intensive to test and maintain. Level A places more stress on response personnel from heat, cold, and psychological standpoints and reduces our visibility and dexterity along with our ease of functioning and performing our mitigation job. We need to select the PPE tools from our Hazmat Tool Box that are appropriate for the circumstances of the hazardous materials release we are called upon to mitigate.

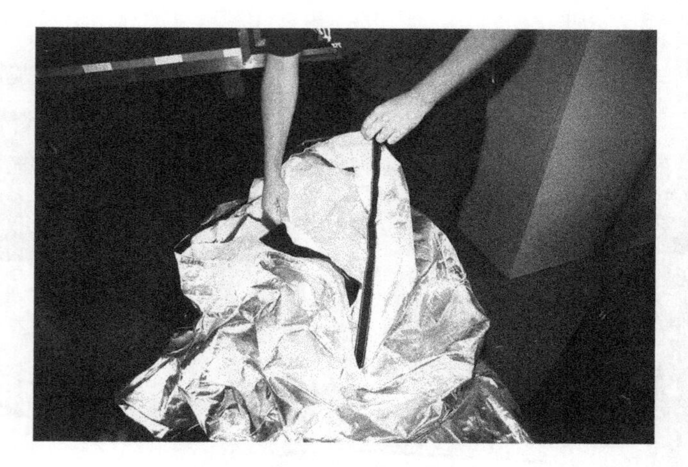

Figure 4.64 Personnel protective equipment (PPE) is another tool in our Hazmat Tool Box.

Risk–Benefit Analysis

Once the incident has been evaluated by first responders and hazmat team members, a thorough risk–benefit analysis needs to be conducted. By conducting this analysis we will be able to make better decisions and determine appropriate objectives. Hazardous materials incidents do not always require actions beyond protecting personnel and the public. There are times when the proper objective would be to withdraw and let the incident take its course without any intervention. This would be done in circumstances where the risk–benefit analysis determines that there is too much risk involved based on the level of benefit that might be gained.

Risk–benefit analysis according to National Fire Protection Association (NFPA) 472 (2008) Edition is as follows: "Risk based response is a systematic process by which responders analyze a problem involving hazmat/WMD; assess the hazards; evaluate the risks; and determine appropriate response actions based on facts; science and the circumstances of the incident." I believe that sums it up pretty well.

Risk Assessment

First, we need to determine the hazard of the hazardous material(s) we are faced with. Hazard evaluation consists of, but is not limited to, physical state, chemical characteristics, physical properties, and other actors. When determining hazard, we also have to look back in history and consider what other responders have done correctly and where there are lessons to be learned by their actions. What is the danger imposed by a particular hazardous material? What other actors are present that may impact the hazardous material? Once we believe we know everything there is to know about this material we can turn to determining what is vulnerable.

Vulnerability

According to Webster's dictionary, vulnerable is derived from the Latin word *vulnus*, which means to wound. Therefore, vulnerability is the capability of being physically or emotionally wounded. In terms of hazardous materials response, this could also be expanded to infrastructure, property, or even the environment. Caution should be exercised here because, if emergency responders are vulnerable and the public is not, we should not place responders in harm's way to protect inanimate objects. It is just that simple. Everything else can be cleaned up or replaced, people cannot, and expending responders should not be an option. Vulnerability then for us should mean who is vulnerable.

Once we determine who is vulnerable, we need to look at how or why. This is where hazards in the hot zone come into play. Those were introduced earlier in this volume. Here, we will talk about them in some detail.

Thermal: Hot or Cold

Thermal hazards that are **hot** generally come from fire or hot materials or objects. Hot materials that are sometimes called elevated temperature materials are usually liquid solids. Molten sulfur, molten aluminum, and asphalt are examples. In New Jersey, in 1949, three men died, two of whom were buried alive under flaming asphalt as a crackling series of explosions destroyed a $500,000 asphalt plant. The shriveled, tar-covered bodies of two volunteer firemen could not be recovered for several hours after they were blown into a ditch filled with boiling asphalt. A third victim, a workman, died of burns later.

Thermal hazards that involve **cold** materials are usually liquefied compressed gases called cryogenic liquids. They are generally not a Department of Transportation (DOT) Hazard Class unless they have some other hazard such as flammability or toxicity. However, the vast majority of them are inert and severely cold however, they are also asphyxiation hazards. Temperatures of up to –450° below zero will freeze and make brittle anything they come in contact with. There is no PPE that will protect your body from these materials. In Springer, OK, in 1998, two workmen froze solid when they became trapped in a liquid nitrogen pipe.

Mechanical

Mechanical hazards usually occur when something falls on you, you get hit by shrapnel, or by a missile thrown from an explosion. Firefighters killed in Carthage, IL and Tidwell, SD were hit by parts of propane tanks that BLEVEed during fires. Enough incidents have happened over the years that we should have learned the dangers of exploding propane tanks. Burning liquefied petroleum gas (LPG) tanks with the relief valve roaring and a flame high in the air is going to BLEVE. We need to just get it in our minds that liquefied gas container fires are losers and vulnerability is extreme. Next to ammonium nitrate, more firefighters have been killed by liquefied compressed gases, including LPG, propane and butane, than any other.

Toxic

Poisonous materials have to find some way to enter your body to cause injury. Responders can fully protect themselves inside the hot zone with the proper PPE. The primary point of entry for poison gases is inhalation, and SCBA can protect against that point of entry. Firefighters should always wear SCBA if they suspect a hazardous material is present. They may not always be visible. Stay out of liquid or solid spills. Before SCBA's were even thought of in 1903, four Milwaukee firefighters, including the

Chief of Department, died a slow death over 24 hours after being exposed to a nitric acid spill they had responded to. They had no idea the acid, which was fairly common, would pose them any harm. Had there been SCBA for them to wear at the time, it would have saved their lives.

Corrosive

Most hazmat teams have Level B PPE that will that will provide full protection from acids and bases, which are corrosive materials. If the acid is fuming, then Level A is required. Corrosive materials can also damage our equipment and may require replacement.

Asphyxiation, Both Chemical and Simple

Simple asphyxiation is just a lack of oxygen. SCBA can protect us from that. Chemical asphyxiation occurs when a toxic material enters the body and prevents oxygen from being taken up at the cellular level. Victims die from oxygen deficiency. Proper chemical PPE will protect responders from both types of asphyxiation. Unfortunately, emergency responders have been killed when they were asphyxiated in confined spaces. The spaces should have been monitored before entry and SCBA worn to protect responders.

Radiation

One nice thing about radiation exposure is you are not likely to encounter it. Containers that are used to transport high-level radioactive materials are so well constructed that I have never heard of a leak occurring from an accident. Low-level radioactive materials are not so well protected, but the hazard from them is not nearly as great. It is still a good practice to conduct a radiological survey of all vehicle incidents just to make sure radioactive materials are present. Low level radioactive materials are often carried in private vehicles or shipped by common carriers.

Etiological (Biological) or Infectious Substance

Etiological or infectious substances are disease-causing living organisms, similar to bloodborne pathogens. Level B PPE should provide adequate protection as most of these materials are solids unless they are in a body fluid (FEMA/NFA).

Consequences

Once again let's turn to Webster's dictionary that defines consequence as "something produced by a cause or necessarily following from a set of conditions." You could apply some physics to this, for every action, there is a reaction. Simply put, if we choose to do something, what is going to happen? Another way to put it, what could happen? Consequence is a result of our chosen actions. Consequences can be favorable or unfavorable.

Likelihood

Webster's dictionary defines likelihood as "the chance that something will happen." Chance implies that it is not certain that it will happen. What evidence can we gather that will determine that there is a good chance something will happen or not happen? Perhaps the best way might be to find out if it happened before by looking at past incidents. We cannot just guess likelihood. There has to be some scientific basis for deciding whether something is likely or unlikely.

Decision Making Process

When making decisions about what to do at a hazardous materials incident, there are three types of data that are necessary to collect before decisions can be made.

Physical

Physical data is gathered at the scene of an incident. Your senses are utilized to search for clues and other signs of what happened, what is happening now, and what may happen in the future. Physical data is necessary before the objectives for the incident can be formulated and other decisions made.

Technical

Technical data comes from a number of sources. CHEMTREC, National Response Center, and computer-based programs such as CAMEO, WISER, and others. Materials Safety Data Sheets (MSDS) and shipping papers also provide technical data. Hard copy reference materials are also a source. Technical data is factual and preverified information that you can count on being accurate.

Cognitive

Cognitive data is what is in your onboard database (the one in your head). Experience, shared knowledge from others, training, and knowledge of previous and historical incidents all contribute to cognitive knowledge (FEMA/NFA).

Recognition Primed Decision Model

Your brain is much like a computer hard drive, it has the capacity of maybe a gazillion bytes of information. When you get involved in any type of incident, your brain scans it and stores the information in your onboard computer. It looks for a match of an incident you have been involved with previously or may have heard about or studied. If the hard drive finds a match, it automatically directs the behavior based on past behavior that ended with a satisfactory result. Simply, the decision-maker has an idea of how things work based on the knowledge that has been gained from experience and study of past incidents. The options are compared against what is known to work.

Historical incidents, like those detailed in volume one, can be substituted for incidents you actually experience yourself. Reading and learning about them will put you there and you can experience what went wrong or what went right. Then, by looking at lessons learned, you can also learn what is known to produce a satisfactory result.

> ***Hazmatology Point:*** *On March 4, 1996, 18 years after the Waverly, TN incident, a similar derailment occurred in Weyuwega, WI. Assistant Chief Jim Baehnman told me when I visited Weyuwega that, "he used his knowledge of the Waverly incident to formulate tactics in Weyuwega." This may have had a direct impact on the Weyuwega incident in terms of safety to emergency personnel and residents. There was not a single serious injury or death as a direct result of the derailment in Weyuwega. Assistant Chief Jim Baehnman said, "from the start of the incident, the tone of the incident would be driven by safety and not time."*

By focusing on the likely rather than the unlikely, you will be able to better handle the incidents you may be faced with, and those skills will also prepare you for the unlikely events that involve "exotic" hazardous materials. Even though you may have exposure to exotic chemicals, that does not mean you will have an incident involving them. By knowing they are in your community or transported through it you can learn about them and can prepare for the unlikely, which will make your response more effective if it does occur. Generally, "exotic" materials are few and far between and incidents are rare (FEMA/NFA).

Traditional Decision Making Models

Before the decision making process can proceed and models selected, all of the data gathering processes must be complete and a mechanism for updated information must be in place. There are several decision-making models available, or you can make up your own, it doesn't matter as long as you go through a modeling process. Three have been selected for this volume. DECIDE and GEBMO were created by Ludwig Benner (1970s) and GEDAPER by David Lesak (1980s). Both of these systems are still valid today.

D.E.C.I.D.E.

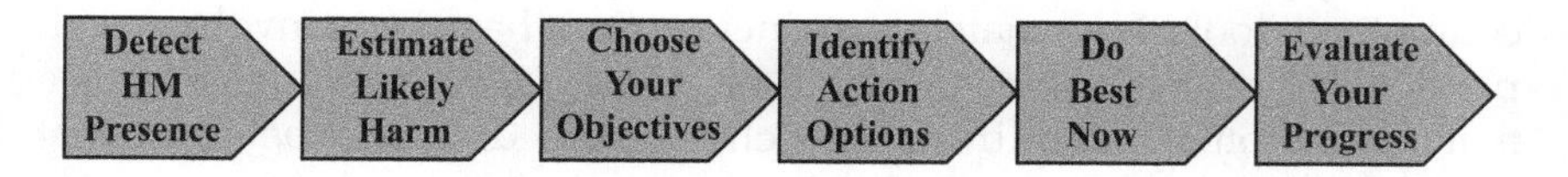

Ludwig Benner "The Father of Modern Hazardous Materials Thinking."

A decisive and innovative change to help firefighters think through a hazmat situation, Benner developed an innovative decision-making process, appropriately named DECIDE.

Detect the Presence of Hazardous Materials

Determining the presence of hazardous materials is the single most important thing an emergency responder can do. It is invaluably important that emergency responders at all levels receive operations-level training and learn how to determine the presence of hazardous materials.

Estimate Likely Harm Without Intervention

This step requires gathering of information about all of the actors in the incident and analyzing the information to determine what is the likely outcome without intervention. Often times this is not a step considered by modern day emergency responders. Most feel they have to intervene because it is part of their job. Emergency responders have been injured and lost their lives over the years when intervention should not have been an option under the circumstances. Without a hazard/risk analysis and looking at harm without intervention, responders may be unduly placed in harm's way.

Choose Response Objectives

Objectives are broad in nature, they need to identify what is the overall incident outcome that we are looking for. Once intervention is deemed necessary, then objectives need to be determined indicating what our overall incident outcomes will be. Objectives do not provide specific information on how the objectives will be accomplished.

Identify Action Options

We need to determine exactly what we want to accomplish and determine if the personnel and resources to accomplish our options are available. Determine what options are available for use based on available personnel and resources. For example, if we have a flammable liquid fire and an option is to put the fire out with foam, it would be necessary to have the amount of foam necessary to put out the volume of fire we are facing. If we do not, then it is not an option.

Do Best Option
Once all of the available options are considered, a decision needs to be made about what option is the best to reach the overall incident objectives. That option or options then need to be implemented.

Evaluate Progress
Determine if the options taken are making progress toward the ultimate mitigation of the incident. If not, then it may be necessary to change options and try another approach and evaluate that. When options are selected without complete information, they may be doomed to fail. Evaluation and implementations of options continue until the overall incident objective is accomplished and the incident is stabilized.

The DECIDE acronym represents key decision-making points that occur during a typical hazmat emergency. "The intent of the DECIDE process," according to Benner, "is to help the responder get 'ahead of the curve' during a hazmat incident." "The goal," he emphasizes, "is to constantly update the predictions of what's going to happen next, in order to see how the actions are changing the outcome. With a hazmat incident, you have to focus on the outcome. The beauty of the DECIDE process is this: If you can't make a prediction about what will happen next, you can pinpoint the data gaps that will ultimately allow you to make a prediction."

Now many years removed from his days of teaching hazmat and investigating incidents for the National Transportation Safety Board (NTSB), but he is still interested in the health and well being of firefighters.

"Back then," Benner pointed out, "firefighters received hazmat training pretty much the same way, following the prevalent fire-service paradigm at the time: attack and extinguish. I wanted to change that paradigm by teaching firefighters the *importance* of thinking their way through an incident rather than jumping into the middle of something they didn't really understand. I wanted to show them how to look at a situation, interpret the visual cues, and predict what was going to happen next."

"Additionally," he continued, "my training program illustrated how critical it is to start out with a game plan, even if it's pretty basic. If the situation isn't going to create a problem, maybe you don't have to do anything. On the other hand, if it's going to hurt somebody, you have to figure out *how* it's going to hurt them and decide whether or not you can do anything about that."

Benner offers this perspective: "I had the very distinct advantage of hindsight when I was investigating accidents for the NTSB, and once you start to understand why people are doing things, you start to see what's going wrong. Back then, firefighters were using the same paradigm for hazmat incidents as they were for structural firefighting – and it wasn't working. All I did was show them how to look at the situation a little differently (by using the DECIDE model) and appreciate the differences between a firefighting mindset and a hazmat mindset. Hazmat incidents can't be handled with a cookbook approach, and I'm not a believer of teaching cookbook-type hazmat training – you have to use your head."

GEMBO

GEMBO is short for **GE**neral Hazardous **M**aterial **B**ehavior **M**odel (Figure 4.65). This model allows an orderly assessment of events that are likely to take place when a container of hazardous materials is stressed. Hazardous materials containers, in general, are designed to hold hazardous material under normal conditions of shipping, storage, and use. When containers are stressed beyond their normal capacity to hold the hazardous material the container is likely to fail. When containers fail certain predictable events will occur, both with the container and the hazardous material. Events may have an unfavorable impact on the public and emergency responders. Impacts may be prevented by removing the public and emergency responders out of harm's way and letting the event take its course. Ludwig Benner Jr. told me this is a time to "Go sit on a hill and watch it happen, you will never see another like it." There is nothing wrong with that choice of action if it is the only safe option. However, using the GEMBO model you can scientifically determine what actions should be taken based upon the incident circumstances when you arrive.

Anywhere on the model where it mentions exposure, it is referring to emergency responders primarily, but it also extends to the public. What you want to do is change the potential outcome, which is what is likely to happen without intervention. The entire purpose of this model

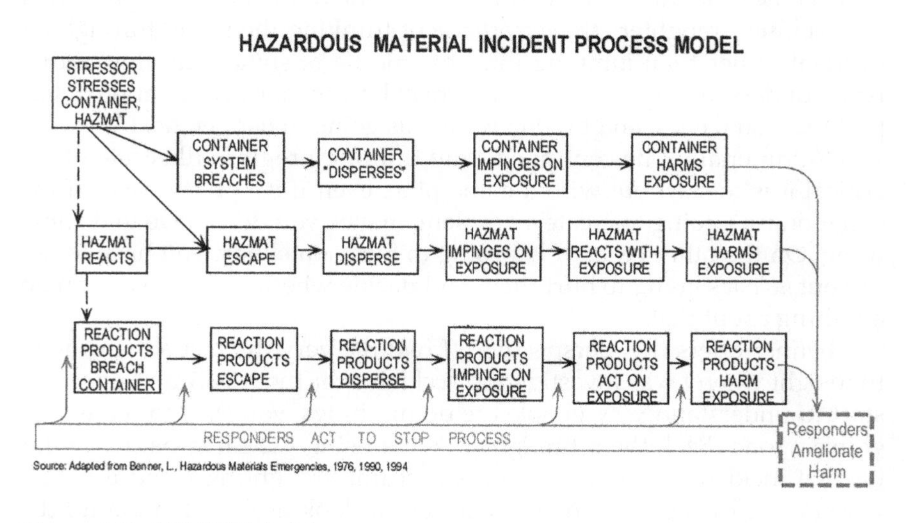

Figure 4.65 GEMBO Model.

is to save lives of responders and the public. Property can be replaced and the environment can be cleaned up, lives cannot be replaced. This model allows you to determine appropriate intervention points during the incident, which is dependent on what is happening when you arrive. History has shown that effects of stressors on containers do not always cause container failure before you arrive. GEMBO provides two courses to take based upon container stressors, container breach or hazmat reactions, which may lead to container breach. When you arrive, either the container is breached or it is not. If not, you may be able to intervene at certain points, again based on when you arrive.

Some of Ludwig Benner's conclusions from incident investigations for the NTSB that led to the development of his models are listed below:

- Traditional attack and extinguishment approaches did not work for hazmat emergencies.
- Firefighters were "programmed" to get into trouble at hazmat emergencies.
- Cookbook approaches to hazmat emergencies produce bad outcomes.
- There must be a better way of responding to hazmat emergencies.

According to Benner, there are two types of responses: adaptive and habitual (cookbook). GEBMO allows for an adaptive approach to an incident based on risk analysis, previous events, and appropriate intervention points. Historically, the vast majority of firefighter deaths and injuries at hazmat emergencies happened before the advent of decision models. That is not to say we have not lost lives since, however, we need to keep educating hazmat responders so that we do not fall into complacency. Dealing with hazmat emergencies can be conducted safely and effectively if we are able to adapt to the situations we encounter and use models (Ludwig Benner).

GEDAPER

GEDAPER is a very similar to DECIDE but illustrates the cyclic decision-making process. Basically, information that is fed into the decision model is the same. GEBMO is a hazardous materials behavior model to figure out potential course and harm to determine objectives. Responders need to work with each or one of their own to determine which works best for them. What is important is that a decision model is used. None of these models will be of much help at an incident if responders are not familiar with them and have used them in tabletop or full-scale exercises. Another way to become familiar with the models is to apply them to historical incidents. You can see if the incident outcome may have been different if models had been available and were used when these incidents occurred (David Lesak).

Personal Protective Equipment (PPE)

Most of us have a wardrobe of clothing that we wear for many different reasons. Some of which are work, comfort, special events, levels of etiquette, weather, and just because we want to. Fire, rescue, and emergency medical services (EMS) have protocols for when certain types of protection from their wardrobe are worn for the hazard being faced. We try to balance protection and function. You wouldn't wear Level A to treat a patient on an EMS call for bloodborne pathogen protection just because you can. You wouldn't wear chemical protective clothing to extricate victims or fight fire just because you can. When fighting fire we do not use a deck gun or master stream on a small fire. We d o not use the jaws of life to cut seat belts. Level A for all entries is like using the deck gun for all fires and the jaws of life for all extrications. Levels A, B, and C PPE have their purpose, just like tools in a toolbox (Figure 4.66). Why should we use Level A on an incident when Level B or even C would work just fine? Use the right PPE for the right hazard. Not one for all, this is not the Three Musketeers.

Bulletproof Vests

On April 20, 1999, two students at Columbine High School in Littleton, CO went on a well-planned shooting spree killing 13 fellow students and teachers. What was fairly uncommon at the time has become a regular event today in emergency response. Not just schools, but practically anywhere else. Motives vary from revenge, political, terrorism, to mental illness. It has become so common today that many departments have come to providing bulletproof vests for fire, EMS, and hazmat emergency responders (Figure 4.67). That is a decision that will have to be made at each local level based upon projected need, cost-effectiveness, and willingness of responders to wear them with all of the other PPE that is required.

Figure 4.66 Levels A, B, and C PPE all have their purpose, just like different tools in a toolbox. (Courtesy: Houston Fire Department.)

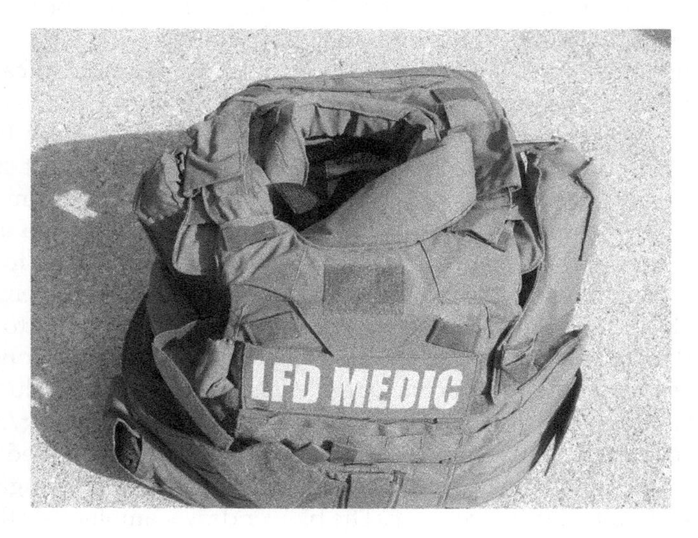

Figure 4.67 This is the type of bulletproof vests used by the Louisville, KY Fire Department.

Rapid Recon/Rescue in a Chemical Agent Environment the 3/30 Rule

Testing done at Aberdeen Proving Ground, MD has been conducted on firefighter protective clothing with military chemical agents. These tests have shown that firefighter turnouts provide a high level of protection should they wander unknowingly into a chemical agent. This testing also led to the development of the 3/30 Rule. Firefighters' turnouts would provide protection to personnel for 3 minutes in a chemical agent environment if upon entry no one is alive in the area. Turnouts would provide protection to personnel for 30 minutes if upon entry people are alive. I do not endorse this procedure. It is provided here for informational purposes only. Each department needs to look at the study and recommendations, do background research and determine if this is something they want to develop into a departmental SOP (IAFF).

Military Agent PPE

Mustard and nerve agents are chemicals and require an appropriate level of chemical protective clothing. Military PPE for battlefield protection against chemical agents is composed of a charcoal suit, protective hood made of butyl rubber protected cloth, butyl rubber gloves with thin cotton inserts, vinyl boots, and a powered air-purifying respirator (PAPR) (Figure 4.68). PAPR is used instead of SCBA because of the length of time that the respiratory protection may be needed, and the difficulty in changing SCBA bottles in a hazardous atmosphere. PAPR removes up to 0.5 mg/m^3 of nerve agent GB for up to 16h based on the longest time an emergency responder has used the PAPR (tests indicate that the filters and cartridges have the shortest service life against nerve agent GB compared with other lethal chemical agents). The actual expected time of cartridge usage during an incident is up to 12h. One of the major problems that occur with any type of chemical protective clothing is heat stress. Because of the extended time the PAPR can be worn, the danger of heat stress is great. In reality, responders will be limited in the amount of time spent in the PPE by heat stress potential rather than the limitations of air supply. Most chemical agents require warm temperatures for them to pose a vapor problem; thus, if there is vapor, heat stress will be a problem.

The PAPR should not be used in immediately dangerous to life or health (IDLH) atmospheres or in atmospheres where oxygen concentration is less than 16.5%. Airborne agent concentration IDLH values have been established for the following nerve agents: GA/GB 0.2 mg/m^3; GD 0.06 mg/m^3; and VX 0.02 mg/m^3. The PAPR uses a battery-operated blower that is designed to deliver essentially decontaminated air at a slight positive pressure into a full-face piece. The blower draws ambient air through two or three air-purifying elements (filters or chemical cartridges), which remove specific contaminants and deliver the subsequent air through a corrugated breathing tube into a face piece assembly on the face of the respirator wearer.

Figure 4.68 Military PPE for battlefield protection against chemical agents.

NIOSH Biological Agent Protective Equipment

The National Institute for Occupational Safety and Health (NIOSH) has released Interim Recommendations for the Selection and Use of Protective Clothing and Respirators Against Biological Agents (Figure 4.69). First and foremost, every agency responding to a suspected biohazard incident should do so with a plan. Elements should include hazard assessment, respiratory protection needs, and decontamination strategies. Plans should be developed based upon recommendations by the Centers for Disease Control and Prevention (CDC) and other recognized expert agencies. Biological agents are particulate materials. They will not penetrate proper respirators and appropriate protective clothing. Based on the following response situations, the NIOSH and CDC recommended protective equipment is listed.

1. NIOSH-approved, pressure-demand SCBA, and full Level A protective suits should be used where the following information is unknown or the event is uncontrolled.
   - Type(s) of airborne agent(s).
   - Dissemination method.

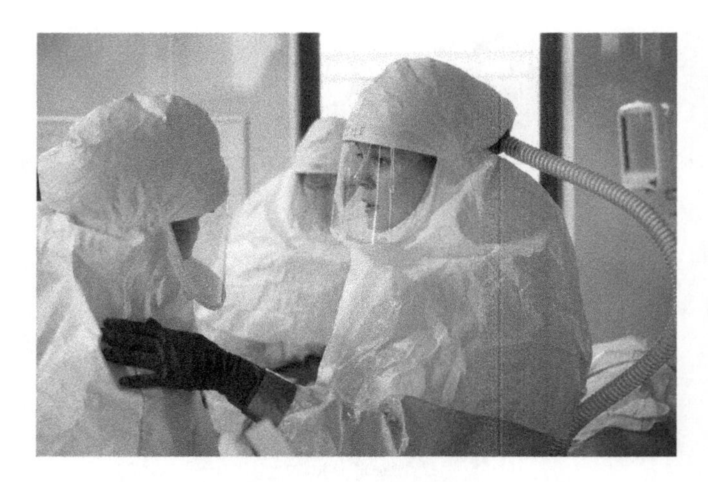

Figure 4.69 NIOSH recently released Interim Recommendations for the Selection and Use of Protective Clothing and Respirators Against Biological Agents, like Ebola.

- If dissemination via an aerosol-generating device is still occurring or it has stopped but there is no information on the duration of dissemination, or what the exposure concentration might be.

NIOSH-approved, pressure-demand SCBA, and full Level B protective suits should be used if it is confirmed that:

- the suspected biological aerosol is no longer being generated, and
- other conditions may present a splash hazard.

Response personnel may use a full-face piece respirator with a P100 filter of PAPR with high-efficiency particulate air (HEPA) filters when it can be confirmed that:

- an aerosol-generating device was not used to create high airborne concentration, and
- dissemination was by a letter or package that can be easily bagged.

Procedures and equipment development are constantly changing, sometimes on a daily basis. It is critical that emergency response personnel keep up with the latest information being disseminated by the FBI, CDC, NIOSH, local and state health departments, and others. If you are unsure about information available, then confirm it with the appropriate organizations. Keep in mind that it is likely that emergency responders will not respond to scenes of biological releases. They will become involved as patients become ill or as investigations of illness develop in an attempt to locate the source of the infection. It is most important that emergency responders and the local medical system are trained to recognize biological agent symptoms and have the proper protective equipment (NIOSH).

Monitoring and Detection Instruments

Monitoring and detection instruments have come a long way over the years. Technology has taken us from paper detection to sophisticated electronic instruments. Monitoring and detection instruments have become more compact and easier to carry and utilize. Because determining the presence of hazardous materials is the most important responder task when they arrive on scene, monitoring and detection equipment can play a vital role in the determination. More and more engine, truck, and rescue companies are routinely carrying monitors on their apparatus. Four and five gas monitors are no longer unusual (Figure 4.70). When the type of monitors are chosen, they should be based upon the most common hazardous materials and the ones most frequently encountered in your jurisdiction.

Monitors are another tool in the Hazmat Tool Box. Don't let sales representatives sell you something they think you need. You determine what you need and tell them. Meters take a lot of maintenance so that they are ready when needed. What you do not need is monitors sitting around that

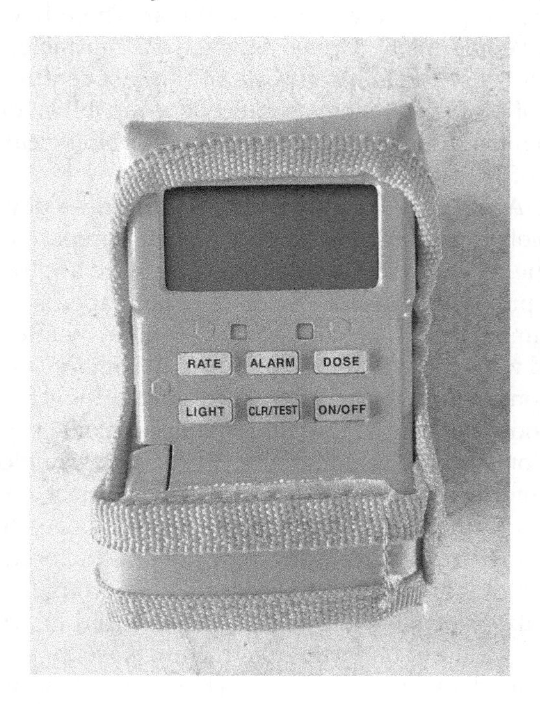

Figure 4.70 Four and five gas monitors are no longer unusual on front line engines and trucks.

you hardly ever use. Over 50% of all hazardous materials incidents involve flammable liquids and gases. Confined spaces generally lack oxygen and may have other dangerous gases inside. Anhydrous ammonia is one of the most common hazardous materials to escape their container. Chlorine is one of the most toxic chemicals you will likely encounter and is very common in most jurisdictions. All of the things just mentioned have a history of being released routinely and killing both firefighters and the public. Those are the things you should be able to detect on a hazmat scene. Radiation, while not commonly released, cannot be detected by any other way than a meter. Radiation meters should also become commonplace on fire apparatus.

Upon arrival at an incident scene, you want to be able to detect LEL, CO, NH_3, Cl_2, lack or excess of O_2, HCN, H_2S, and radiation. When the hazmat team arrives they should be able to detect the same chemicals and have some monitors to detect unknown materials and any other hazardous materials that are common in your community and are frequently released. Some monitors are carried that will likely never be used. Military agent monitors are fine for large metropolitan areas, but there has never been a release of a military chemical agent in the United States by terrorists or accidents. Your chances of winning the lottery are better than encountering military chemical agents. It is really not a credible threat. Biological agents, however, are a threat. There have been attacks with anthrax. People in the United States have contracted Ebola. White powder incidents are the bomb threats of the 21st century. Detection or identification of biological agents should be a capability of most departments. Not so much to identify anthrax or other biological agents, but to rule them out of white powder incidents.

Anniston, AL Center for Domestic Preparedness developed a very inexpensive biological detection kit that you can prepare at your department. Just some common items from the drug store and other local businesses can be put together for a few dollars. pH paper is inexpensive but can give you important information about what a chemical or biological material is and what it is not. Reno, NV hazmat team for years has focused on identification by monitoring and have reduced their time on scene by almost half. Knowing what a hazardous material is, or at least its hazard is, tells what our options are in mitigating the hazard. Using common sense we may find intervention isn't always necessary. Detect, contain, and call the contractor to mitigate the hazardous material. Once an emergency is over, our job is just about done as far as hazardous materials incidents go.

Purchasing monitors should be based upon your hazard needs in your community. Expensive and not frequently used instruments can be purchased by a group of departments that are close by to avoid duplication. Talk to other departments and see what brands have worked best for them before you make your decisions. Just make sure your Hazmat Tool Box has the monitoring tools you need and know where to get additional resources on the rare occasions you might need them.

Types of Container Failure

Information here on types of explosions can be a confusing and controversial subject. Some of this information presented is my opinion based on my research on what has happened at hazardous materials incidents. It does not necessarily agree with commonly accepted theory. There are three types of explosions that can occur involving containers, boiler explosion, container rupture, (called Heat Induced Tear HIT in the 2020 ERG) and BLEVE. Boiler explosions occur, of course, in boilers, but what we are really talking about here is a boiler type or similar explosion. For example, water is used in shipments of phosphorus to keep the air reactive material from reaching air in transportation, use and storage. Phosphorus is shipped in liquid containers. Incidents have occurred where water has escaped a container allowing phosphorus to catch fire. The fire has heated the remaining water in the container changing it to steam, a gas. Liquid containers are not designed to contain gases. Explosions occurred from the over pressurization of the container caused by the steam. This is a boiler type explosion. Boiler explosions do not involve flammable liquids or gases.

B.L.E.V.E.

This acronym stands for the following phenomenon: **B**oiling **L**iquid, **E**xpanding **V**apor, **E**xplosion. BLEVE is a violent explosion and tank failure that breaks the tank into pieces that are rocked several thousands of feet away from the blast center (Figure 4.71). BLEVE is the second leading cause of death to emergency responders dealing with hazardous materials. This phenomenon in my opinion only applies to liquefied compressed gases shipped, stored, and transported in pressure containers where the liquid exists in the container above its boiling point. If you look at the

Figure 4.71 BLEVE is a violent explosion and tank failure that breaks the tank into pieces that are rocked several thousands of feet away from the blast center.

explanation of BLEVE as it was originally put forth it really explains the concept. Boiling liquid, the liquid in the container is there above its boiling point and kept in the container by pressure. The boiling liquid is a liquefied compressed gas, not a typical liquid. Expanding Vapor, when the pressure is released from the container all of the liquefied gas instantly returns to the gas state. No liquid is left in the container. This process occurs explosively. Fire may occur or a fireball if the gas released is a flammable gas and it reaches an ignition source. Hazards that are presented by this scenario are much different and more severe than a fire and release of product involving a liquid container. We are talking about common hazardous materials like propane, butane, and LPG in terms of flammability. Nonflammable liquefied containers can also BLEVE, such as anhydrous ammonia, chlorine, and others. These are not "exotic" materials and are found in most communities.

> ***Author's Note:*** *During the 1970s an anhydrous ammonia tank that was being filled at the Port of Catoosa near Tulsa, OK BLEVEed, killing two workers and spreading ammonia downwind, defoliating trees, and turning vegetation brown. The cause was a bad weld during a repair of the tank in South America.*

Liquefied compressed gas is maintained as a liquid by pressure in the container and the construction of the tank itself. Liquid does not fill the entire space inside the tank, there is a vapor space above the liquid space. Temperatures of these liquids are ambient. Whatever the ambient temperature around the containers is the temperature of the material inside the container will be the same or nearly the same. Radiant heat from the sun or exposure can raise this temperature.

When an accident occurs that causes a fire to impinge upon a liquefied gas container, an increase in pressure within the tank will occur. When the impingement occurs on the liquid level, the liquid temperature will increase, causing the liquid to boil faster and release more vapor, creating more pressure in the tank. Liquid in the tank actually absorbs the heat, protecting the integrity of the tank shell; thus, prolonging the integrity of the tank shell. Increased pressure will activate the pressure relief valve, releasing some of the pressure build-up. Rapid boiling will also reduce the liquid level of the tank. Relief valve function may be able to handle the pressure increases as long as the tank remains intact. Responders should quickly place unmanned water streams in place to aid in tank cooling, and responders should evacuate immediately to a safe distance. Incidents have occurred where tanks did not BLEVE and the fuel in the tanks vented until near empty.

However, if flame impingement is on the vapor space of the tank, it is a true life-threatening emergency. Actions to protect responders and the public should be immediately undertaken to save lives and injuries.

NFPA says BLEVE will occur within 8–30 minutes with an average time of 15 minutes. Fifty-eight percent of incidents that have occurred in the past happened in 15 minutes or less. Do we know when flame impingement began? How long did it take us to get there? How long will it take us to set up cooling streams? How high is the flame above the tank? How loud is the roar? Has it changed since we arrived? Just food for thought, but we do not have much time to think or consider tactical options. This should be one of those take out the tennis shoes and run as fast as you can scenarios. Flame impingement on the vapor space is a loser from the start. Evacuate all that you can, withdraw, and let whatever is going to happen with you nowhere near it.

Container Rupture

Rupture of containers occurs with liquid or atmospheric pressure containers (Figure 4.72). No gases are shipped, stored, or used compressed or are otherwise legally placed in these liquid tanks. Commonly, ethanol, methanol, crude oil, among others are what we expect to find in liquid containers. Over the past decade, numerous train derailments have occurred around the United States and Canada involving ethanol and crude oil. Many of the derailed tanks caught fire, impinging on other tanks, causing them to catch fire. Flame impingement caused many of these tanks to fail, but not violently like a BLEVE. No tanks or pieces of tanks were rocketed. Fireballs occurred but were from burning vapor that was released from the tanks. All of the liquid did not instantaneously turn into a gas or vapor. Tanks continued to burn because there was still flammable liquid left in the tanks to burn. No further ruptures or explosions have occurred to my knowledge once a tank opens up. Initial pressure is released and

Figure 4.72 Rupture of containers occurs with liquid or atmospheric pressure containers. (Courtesy: Franklin County, Maine, USA.)

any further creation of pressure will be vented through the first rupture hole. When incidents were over the intact tanks remained with obvious tears or ruptures in the tank wall.

These tanks ruptured, they were not boiler explosions or BLEVEs. If you look at photos of railcars following fires with ethanol and crude oil, they are usually lined up next to each other and you can see the holes in the tanks where the ruptures occurred. Otherwise, the tanks are all intact. To my knowledge, no firefighters have been killed battling fires in ethanol or crude oil tanks. Deaths to civilians have occurred but not in all incidents. Look in the emergency response guidebook (ERG) for evacuation distances for ethanol, crude oil, and propane. Ethanol and crude oil railcar on fire evacuate 1/2 mile. For propane, this distance is 1 mile. If liquid and pressure tanks both BLEVEed, the distances would be the same and history has shown they have not been. Ethanol and crude oil liquid tanks rupture. Liquefied compressed gas pressure tanks BLEVE.

Boiler-Type Explosion

Phosphorus is a flammable solid that reacts with air above 86°F and spontaneously combusts upon contact. To safely transport phosphorous, water is placed in the container to keep it from reaching air, which is the ignition source for the material. Phosphorus tank cars are liquid cars. This scenario works fine as long as the container safely travels during transport. However, if a derailment occurs involving phosphorus and the container is breached or the tank opening is damaged and water leaks off, a fire will occur. If any water remains in the tank, the burning phosphorus will heat the water to its boiling point and turn it into steam. Now you have a gas-forming inside of a liquid car, and when pressure from the gas builds up to a point the container can no longer hold it, the tank, not the phosphorus, will explode. This is a boiler-type explosion.

These incidents actually happened in Brownson, NE on April 2, 1978 and Miamisburg, OH on July 8, 1986. Outcomes were totally opposite. In Brownson, the tank exploded but in Miamisburg the tank remained intact. Both tanks tipped on their side during the derailment allowing water to spill from the tanks and the phosphorus ignited in both cases. Miamisburg's tank car was much more severely damaged with various punctures on the top of the tank. It is my opinion that the steam buildup inside the Miamisburg tank was vented though these puncture holes and did not build-up to the point of tank failure.

The same thing can happen with 55-gallon drums in a box truck. If one develops a leak, the water will run off and the phosphorus will ignite. The heat from the first barrel failure will heat the water in the remaining barrels, and they will fail one or several at a time until all of the drums have exploded and the phosphorus is consumed. This is also a boiler-type explosion. Tank car parts or barrels may rocket several hundred feet from

Figure 4.73 This incident occurred in Gettysburg, PA on March 22, 1979. (Courtesy: Gettysburg, PA Fire Department.)

their original location. No fireball will develop because the steam is not flammable. An example is the incident in Gettysburg, PA on March 22, 1979 (Figure 4.73).

Flammable Liquid Fire Phenomena

Slopover, Frothover, and Boilover Phenomena of slopover, frothover, and boilover need to be defined. A slopover results when a water stream is applied to the hot surface of burning oil, causing the burning oil to slop over the tank sides. A frothover is the overflowing of a container not on fire when water boils under the surface of viscous hot oil. An example is hot asphalt loaded onto a tank containing some water. The water may become heated and start to boil, causing the asphalt to overflow the tank.

Compare these events with the definition of a boilover: a sudden and violent ejection of crude oil (or other liquids) from the tank resulting from a reaction of the hot layer and the accumulation of water at the bottom of the tank. A boilover occurs when the residues (heavier particles remaining after combustion) from the burning surface become denser than the surrounding, less dense oil, and the residues sink down below the surface level toward the bottom of the tank. As the hot layer of more dense, burned oil moves downward, this "heat-wave" will eventually reach the water that normally accumulates at the bottom of a tank. When the two meet, the water is superheated, and subsequently boils and expands explosively, causing a violent ejection of the tank contents.

Although the normal water-to-steam expansion ratio is 1,700:1, this is at 212°F. At higher temperatures, the water-to-steam expansion ratio can be as much as 2,300:1 at 500°F. If a boilover occurs, a rule-of-thumb says that the expelled crude oil may travel up to ten times the tank diameter around the tank perimeter. For example, in a crude oil tank 250 ft in diameter, expect crude oil to cover an area of 2,500 ft from the tank. These figures have never been tested for very large atmospheric storage tanks of 300 ft or greater, therefore in tanks of this size, the distance of ten times the tank diameter may have to be increased.

Therefore, carefully consider the location of the incident command post, staging, equipment placement, medical triage, and a safe zone. Identify and use additional firefighting resources. Local plant fire brigade members can provide the much-needed support and technical advice. Many of the hardware resources may be available at the facility or through industrial and municipal mutual-aid agreements. Some facilities have contracts with private third-party companies that specialize in extinguishing large hydrocarbon fires (FEMA/NFA).

Decontamination

Common Sense Decontamination

While gathering updated information on Allegheny County, PA Green Team for inclusion in Hazmatology, I learned about their innovative common sense approach to decontamination (Figure 4.74).

> ***Author's Note:*** *While preparing for the Hazmatology book project, decontamination was one of the areas that I felt science would back up the idea of taking another look at the way we approach decontamination. Finding a team that was already using "common sense" for decontamination was very refreshing and worth sharing.*

Green Team created a flow chart to assist them in determining what level of decon is needed based upon the circumstances of an incident. Decon is another example of a tool in the Hazmat Tool Box that is not always needed. Using tools based upon need rather than because we can or think we should only adds additional tasks and stresses that are not necessary. Entering the hot zone in Level A PPE for gases does not result in contamination of the chemical suit. Gases do not stick to a surface. While in the hot zone, if personnel do not come in contact with a hazardous material, there is no contamination. Generally, we can see the hazardous material if it is a liquid or solid. If a solid is not suspended in air and does not come in contact with PPE, there is no contamination. Therefore, no decontamination is required.

Allegheny County, PA Green Team's Common Sense Decontamination Philosophy and Concept

Introduction If you were to look at the meters and air monitors that were available to hazardous materials responders in the 1980s during the dawn of hazmat response and compared them to today's world of 4 gas meters with integral PIDs, infrared spectrometers and even truly portable gas chromatographs, the advances in just 30 or so years are astounding. Much

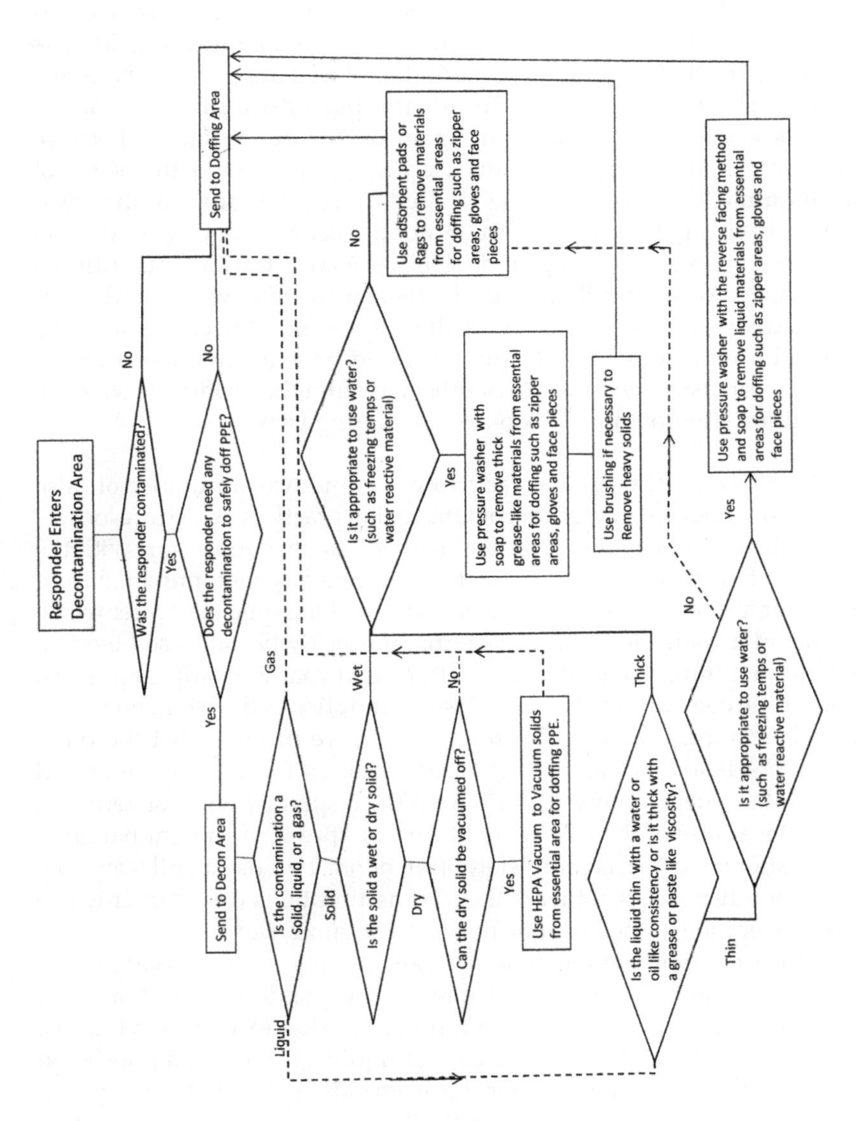

Figure 4.74 Innovative common sense approach to decontamination.

of the same can be said for the PPE and information technology available to responders. In spite of that, the hazmat community is still teaching new technicians to decontaminate in the same way our predecessors did 30 years ago.

After nearly 20 years in the hazardous materials business, we decided there has to be a better way to decontaminate people in the field. The impetus for this thought was an episode that played out during a National Pro Board Hazardous Material Technician practical exam. One of the evaluators wanted to fail the entire team on the decon station because some of the simulated contaminant used was found under the SCBA of the contaminated responder being cleaned. It was then that we first realized that we had gotten the whole concept of decon all wrong. Two years of deliberating, experimenting, and trials has lead to the method outlined in this paper. A method that is more efficient in time, water, and manpower without risking the safety of the responder. Our colleagues and us poured a lot of our life into this idea and have accepted and evaluated much criticism, constructive or otherwise and the product is what we believe decon should and will look in the coming years.

False Assumptions The first obstacle to overcome was a number of false assumptions about our current decontamination practices, which are loosely based on the EPA decon methods. For instance, many people start with the assumption that decon means completely cleaning any contamination from a responder at a technical decon line in the field. Once you start attempting to do this in a controlled training setting, it quickly becomes self-evident that totally cleaning someone is a fool's errand that is nearly impossible and very time consuming. Another close assumption is that someone's PPE *needs* to be completely clean. On the contrary, we propose that the outer PPE only needs to be clean enough. Clean enough for the responder and an assistant to safely remove the PPE from the responder without getting it on him. Once clear of the PPE without further spread of contamination, it doesn't matter if the PPE is completely clean or not, because it will simply be packaged for disposal. Another of the false assumptions we commonly live by is that everyone in the hot zone needs decontaminated.

Again, after applying much thought and good science, it is easy to see that this is just not the case. For instance, if responders were in a cloud of gas or vapor, such as chlorine or ammonia, what exactly is it that we would be decontaminating? In the case of liquids, if there is a puddle on the ground and it is obvious no one has been close to it, let alone stepping in it, do they need deconned? And lastly, if someone does step in a puddle of diesel fuel, would you really need to decontaminate their whole body or just their feet? As we dissected some of the currently accepted methods of decon, more and more sacred truths began to unravel under the scrutiny of logic and science.

Why would we want to fix something that already works? Decontamination as it is popularly practiced in the country today is a resource intensive operation. When examined closely, is it really providing value commensurate with the resources it exhausts? For instance, if you set up a basic 3 step decon, it can involve anywhere from 3 to 7 people, gallons of water per person deconned and a plethora of hardware store "junk" including buckets, tarps, hoses, brushes, pools and maybe even a portable shower. In addition to the amount of resources consumed, more taxing can be the time involved. Often, decon lines at larger incidents can be seen backing up and low air alarms can be heard ringing as decon teams race to move responders through the line in time to get everyone safely out of their PPE. Our method of decon both dramatically reduces the resources used and the amount of time it takes to decon responders. Using this method properly, a responder can be decontaminated and out of their PPE in under 3 minutes all while using only about 1 gallon of water per responder.

So how does it work? The Common Sense Decon method works in 3 basic parts. The first is a decision making flow chart that allows the decon team leader to quickly make important decisions for providing the most effective decontamination. This flow chart directs the decon decision maker in selecting whether or not to perform decon at all, whether a wet or dry method should be used, and when to send someone to doff their PPE.

The second part is the actual method for removing contamination. For situations involving dry solids, such as dust and powders, a HEPA vacuum is used as the primary means of removal. For wet solids or liquids, with consistencies from thin oils to heavy greases or sludge's, a lower pressure electric pressure washer is used. In either case, only enough contamination is removed to allow the responder to safely remove their PPE. To spend time removing any more contamination is simply wasteful. Our experiments show that cleaning the hands and the zipper areas of suits or turnout gear is the most important, but on an individual basis, a determination can be made as to whether boots or heads may also need to be cleaned. The actual methods recommended are detailed in the section entitled "Removing Contamination".

The third part of the Common Sense Decon method involves the doffing of PPE. The basic method for chemical protective clothing is to start by removing any tape from around the gloves (for level B and C) and then opening the zipper. The area around the zipper should be clean from the previous station. Next, move to the top (after removing SCBA when in Level B) and begin rolling the inside out and down and over the shoulders trapping any remaining contamination inside as the suit is rolled down. As the suit is rolled to the shoulders, have the responder make a fist to capture their gloves and pull each sleeve inside out, letting go of the gloves after the sleeve is inside out, but in time to allow the glove to remain in the sleeve. Then, continue rolling the suit down to the feet and over the boots, leaving the rolled up suit around the boots, like turn-out pants pulled down around fire boots.

From here, the responder can step out of the boots and onto a clean adsorbent pad (or other clean surface). The decon team member assisting with decon can lift the boots and rolled down suit as a single unit and throw it into a waste drum. In another possible variation, before doffing, the responder can actually step unto a rolled down 55 or 85 gallon drum liner (Trash Bag) so that when they step out of their boots, the decon team member can simply close the bag, tie it, and then lift the whole bag into a waste disposal drum.

It is that simple. In testing, it proved consistently to work in a safe and effective manner and was able to push people through the decon line in about 2 minutes. In all tests, responders were able to remove their PPE with assistance from a decon team member without spreading contamination to themselves. The bottle neck in the system is the doffing station which accounts for about 75% of the time. The pressure washer can clean a responder in about 30 seconds using the procedure outlined in the "Removing Contamination" section. The system can be effectively run with just a washer and a PPE doffer, but to overcome the bottle neck caused by the doffing, we recommend one washer with two separate doffing stations accepting people alternately for doffing of PPE.

Removing Contamination The saying, "The devil is in the details" is as relevant here as anywhere. To effectively remove contamination, there are several details that come into play. The first is the use of a pressure washer. After hearing all of the possible cons to this plan that could be dreamed up, we just decided to try it. We conducted experimentation involving turn out gear and chemical protective clothing. As contaminants we used everything from peanut butter to axle grease. Those experiments set us straight on a number of details. First, a pressure washer can be safely used on a person if the pressure is fixed low enough and the spray pattern is also permanently fixed at an acceptable angle.

We decided on an electric pressure washer at around 1200psi which had interchangeable tips (hard to find on electric models) as opposed to the adjustable single tip. We decided that by selecting a tip with an appropriate angle such as 30° we wouldn't have to worry about someone mistakenly using a thin concentrated stream, which will damage suits and possibly injure people. In addition, we looked for a model that could induct soap right from an onboard tank without switching wands or nozzles.

The key to successfully and safely pressure washing a person whether in turn-out gear or chemical protective clothing is the method. This is one of the areas where specific training matters. The method we developed is not what we are traditionally trained to do. This method starts by pressure washing the area around the zipper (about 6″ on each side). The pressure washer should hold the gun high and point it at a downward angle. Starting at the top of the zipper, the pressure washer should work his way down the zipper in short up and down strokes, pushing any visible contamination downward. Once complete a decision must be made. If the contaminated responder is in Level A, continue by pressure washing the hands and possibly the boots if heavy visible contamination exists there.

If the responder is wearing anything besides Level A, have them turn around and face rearward, extending their hands out in front of them. This allows for two critical things. If the suit is worn in the traditional way with the sleeves over top of the gloves, it allows the pressure washer to push contamination down the glove towards the finger and off the end without blowing it up under the sleeve, which could happen easily with a pressure washer despite the presence of tape. The other advantage of the deconee standing backwards is that material can be blown off of the head if necessary without forcing it between the face piece and the hood. This was especially important on our team since we don't tape around face pieces. Our belief is that if you need to tape around a face piece, you need a different suit, which has led to the integration of encapsulated CPF3 or similar suits for some Level B work where heavy contamination, especially around the head, is likely.

Pressure washers will fling material and make a cloud of mist, so we decided we would need a backstop to collect the water and material that was coming off our deconee. The solution that we chose after a great deal of trial and error was a three sided structure built from pvc pipe with a removable plastic cover. The backstop was built with commonly available materials from the local home improvement store and the plastic covers can be made for a relatively low cost and are easily removable for disposal. Although this problem could be solved an infinite number of ways by teams wishing to use the method, we chose a method that allowed for quick assembly by even those with little training and allowed for easy storage on our apparatus.

Another paradigm shift was our approach to collecting the decon water. In most cases…we are not. Looking at the responses by the 5 teams in our county over the last 10 years we determined that we just are not getting that contaminated.

Using our method is creating very little waste water and in every case where a wet decon was used in the past, environmental clean-up contractors were later called in to remediate the scene. At about 1 gallon per person deconed, we decided that the simplest answer was to just let the run-off water lie on the ground. The area contaminated by the water will be insignificant and can be remediated by the clean-up contractor if necessary. We also understand that there will be exceptions to this, in which case we will still carry one collapsible decon pool for water collection, although we don't anticipate needing it.

It is likely that few people would argue the need to improve the current decontamination systems in use within the world of hazardous materials response. The current methods are resource intensive including manpower, equipment, and training time. Worst, all of the resource drain typically leads to results that are marginal at best. Our Common Sense Decon method reduces the resource load while improving the efficiency of the decon process. 2014 will be the field testing period for this method within our team following months of development and experimentation. We hope this method will gain widespread acceptance and give teams everywhere a viable alternative to what they are currently doing or at least point them in the direction for making their own needed improvements to the system. When looking at the advances that have been made in the technology we use today, it becomes easy to see the need for updating decontamination processes as well so that all parts of our response operations remain progressive for the safety of the public and our responders.

Dressing-Out Procedure – Level B Coveralls:

1. Don the Coveralls
2. Place inner and outer gloves on and tape the otter gloves to the suit using a 1″ tap.
3. Placing booties into boots, DO NOT APPLY TAPE TO THE PANTS LEG AND BOOT.
4. Don SCBA including mask
5. Place hood over head. DO NOT TAPE HOOD TO MASK.
6. Taping front zipper is CONDITIONS SPECFIC based upon the likely hood of becoming contaminated.

In conditions where there is a likely hood of the product becoming in contact with the suit, determine if the coveralls type ensemble in the best choice for that environment.

Decon – Wash

A. Establish Decon area in the WARM ZONE
   1. Traffic Cones Identifying the Area and the Entrance
B. Establish DECON DRESS DOWN AREA in the COLD ZONE
C. Setup 3-sided partition
D. Setup Water Heater
E. Place Blue "A" bucket (5 gal) with brushes, scrapers and wipes near partition with easy access for Washer
F. Fill another bucket (5 gal) with soap/water solution with easy access for Washer
G. Place non-skid mat and/or hydrophobic pads on the ground in the 3-sided partition for the Deconee to stand on.

H. Establish water to Decon area using garden hose; Use manifold with 1 1/2" hose line when necessary.
I. Establish power source at 110v AC for Pressure Washer
J. Connect Power Cord and Water Line to Pressure Washer

Estimated Time to accomplish: 15 to 30 seconds

1. Place the Deconee inside the 3-sided partition facing the back wall
2. Have the Deconee raise their arms in front of their body; perpendicular to their body
   ALL PRESSURE WASHING SHOULD BE DONE AT A DISTANCE AS NOT TO TEAR THE ENSUMBLE, *APPROXIMATELY 1 to 2 Feet*, but close enough to clean the suit material and gloves.
3. PW the gloves
4. PW the front of suit so that Deconee can make a CLEAN EXIT from the suit
5. VERIFY DECONEE IS CLEAN AT EGRESS POINTS

Decon – Dress down

This Process Requires at Least 2 Dress down Areas

A. Establish Dirty and Clean Side of the Dress down Area
B. Place 2-Step Ladders in Dress down Area
C. Place Large Garbage Bag (60 gal) in Area so that Deconee can step into bag with both feet at the same time
D. Prepare clean clothing to be donned by Deconee including Foot Protection.
E. Prepare to place pad(s) in the ground where Deconee will step after suit is completely doffed and feet are removed from boots.
   1. Remove MMR from face piece while keeping face piece on Deconee's face.
   2. Remove SCBA
   3. Start at hood by rolling the suit inside-out to the shoulders
   4. Remove the suit off of the right shoulder
   5. Remove the right arm from the sleeve while turning the sleeve inside-out at the same time by making fist with their hands in the Outer Gloves, and KEEPING THE INNER GLOVES ON THEIR HANDS when removing their hand from the Outer Glove
   6. Remove the suit off of the left shoulder
   7. Remove the left arm from the sleeve while turning the sleeve inside-out at the same time but keeping the INNER GLOVES ON THEIR HANDS
   8. Direct the Deconee place his arms crossed on his(her) chest and use interlocking thumbs
   9. Continue Rolling the suit down to the top of the boots
   10. Unlatch the boot ties

11. Place the Deconer's foot between the feet of the Deconee on the inside of the boot to assist the Deconee in removing their feet from the boots.
12. Direct Deconee to step out of their boots and onto the pads while holding onto walker
13. Direct Deconee remove their face piece and hand it off to be placed in dirty area for cleaning
14. Remove Deconee's Inner Gloves
15. Direct Deconee to don clean clothes and shoes
16. Send Deconee to Medical Surveillance for evaluation
17. Close the Decon bag with the suit and other items for disposal
18. Replace Standing Pads when necessary for next Deconee
19. Remove disposable items and prepare for the next Deconee

(Allegany County, PA Green Team)

Conventional Decontamination

Decontamination is just another one of the tools in your Hazmat Tool Box. If found to be necessary, decontamination is one of the most important actions taken by emergency personnel during a hazmat or WMD incident (Figure 4.75). Decon reduces the effects of hazardous materials and terrorist agents when response personnel and the public become contaminated. Decon also limits the hazardous materials or WMD agent to the "hot zone" and "warm zone" and prevents cross-contamination.

Decontamination Definitions

National Fire Academy According to the National Fire Academy, "Decontamination is a chemical or physical process used to remove and

Figure 4.75 If found to be necessary, decontamination is one of the most important actions taken by emergency personnel.

prevent the spread of contaminants from an emergency scene due to their ability to cause harm to living beings and/or the environment."

Occupational Safety and Health Organization (OSHA) "Decontamination is the process of removing or neutralizing contaminants that have accumulated on personnel and equipment." This is sometimes referred to as technical decontamination.

Department of Homeland Security (DHS) Decontamination of patients is: "Any process, method, or action that leads to a reduction, removal, neutralization – by partitioning or binding (as opposed to chemical neutralization, which is not recommended; see GS 2.11) – or inactivation of contamination on or in the patient in order to: prevent or mitigate adverse health effects to the patient; protect emergency first responders, health care facility first receivers, and other patients from secondary contamination; and reduce the potential for secondary contamination of response and health care infrastructure."

Centers for Disease Control Decontamination is the reduction or removal of radioactive contamination from a structure, object, or person.

NFPA

The NFPA defines decontamination as the "physical or chemical process of reducing and preventing the spread of contaminates from persons and equipment used at a hazardous materials incident."

Contamination, or the event that makes decon necessary in the first place, occurs in the hot zone. The determination if decontamination is necessary should be based upon the physical state of the hazardous material and if contamination actually occurred.

Three distinct zones are established during a hazmat incident: hot, warm, and cold. Anything or anyone who has been contaminated, including responders and victims and the product and container, are located in the hot zone. The warm zone surrounds the hot zone and it is where decon occurs. It is sometimes referred to as the contamination-reduction corridor. Outside of the hot and warm zones is everything else. No contamination should be present and it should be a reasonably safe area. This area is referred to as the cold zone. There are two primary types of contamination, direct and cross. Direct contamination occurs in the hot zone. Cross-contamination occurs when someone or something outside the hot zone was not properly decontaminated and comes in contact with another object or person, usually in the warm or cold zone. Therefore, the real mission of decontamination is to prevent cross-contamination from occurring in the first place.

Equipment and the environment are secondary concerns. Entry team personnel, even with proper PPE, should not come in contact with hazardous materials in the hot zone unless they absolutely must as part of mitigation efforts. This will help keep contamination to a minimum and make decon easier or decon may not be required at all. Besides the two types of contamination, there are also two primary types of decontamination, technical and emergency.

Technical Decontamination

Technical decontamination occurs when the hazmat team arrives on scene. It may involve decontamination of response personnel only or both responders and victims. When victims require decontamination, two or more decon lines will need to be established, one for personnel and the other for victims. Victims may be ambulatory or may need to be removed by properly protected responders through the decon line. Either specially trained decon personnel or hazmat team members will set up technical decontamination.

Some departments have developed trained decon units that respond when the hazmat team is dispatched on an alarm. They perform only decon functions. In some jurisdictions decon units are also being utilized for firefighters to remove cancer causing contaminants from bunker gear personnel should not go on air supply until the entry team is preparing to exit the hot zone. This is done to ensure that decon team members have enough air supply to perform decontamination on entry personnel and themselves.

Technical decontamination is performed with equipment carried on the hazmat or decon units for just that purpose. Equipment is not highly technical and does not have to be expensive. The process requires a device to catch contaminated water; buckets and hose; tarps to set collection devices on; and soap, scrub brushes, and trash containers. Water-catching devices or pools are available commercially that have steel frames and plastic liners and other types. Children's swimming pools, both rigid and inflatable, also work well, are disposable, and available where toys are sold. Other supplies and equipment can be purchased from home improvement or hardware stores.

Decontamination relies heavily on water, so supply must be established from a domestic source, fire hydrant, or fire apparatus. Volume should be kept to the minimum necessary to accomplish the job as the runoff may need to be contained. Following the incident, environmental or health officials should be consulted for disposition of decon runoff.

Soap and water are the only decon solutions that should be used on victims or personnel in PPE. Bleach should not be used on victims, even if a chemical or biological agent is involved. Bleach can cause injury and even blindness if it gets into the eyes.

Decontamination, typically, has been a part of hazmat team operations. It generally involves the use of a decon corridor (sometimes referred to as contamination-reduction corridor) located in the warm zone with a gross contaminant reduction section followed by a series of wash and rinse stations, an undressing area, and a rehab area. Many departments train and equip specialized decontamination teams with personnel who have operations-level hazmat training and provide them training for decontamination. OSHA's basic regulatory philosophy is to train and equip employees to do the job they are asked to do. Many jurisdictions do not have their own hazardous materials teams. They require the use of mutual aid or private contractors for their hazardous materials response. Mutual-aid teams and private contractors generally do not respond with large numbers of people.

A simple entry into a hazardous materials hot zone and resulting decon at a hazardous materials incident can easily require 6 or more personnel. So, the idea of a local designated decontamination team can provide additional personnel to assist the mutual-aid hazmat team with decontamination. This will free up technician-level personnel on the hazmat team to concentrate on entry operations. The local decontamination team can provide ambulatory victim decontamination while waiting for the hazardous materials team to arrive. The local decon team can then provide technical decontamination for hazmat team members.

Fire departments usually have access to SCBA for respiratory protection. OSHA regulations permit decontamination teams to wear one level of protective clothing less than the entry team is wearing. Hazmat teams generally use Level A (total encapsulation) for vapor atmospheres or Level B (splash protection) for nonvapor atmospheres. As a result, the decontamination team will almost always be in Level B protection. Decon personal should not wear firefighter turnouts to conduct decon. Disposable Level B suits are not that expensive. As I said earlier, a decontamination team can be outfitted and equipped for a modest expenditure of funds.

Training has a monetary value in terms of the cost of the training and in the cost of the time spent on the training. Typical decon training might require 8–16h which would include theory and hands-on. Training should include, some elementary toxicology, decontamination solutions, decontamination setup, protective equipment, and decontamination procedures. Toxicology is covered because the primary reason for decontamination is to remove toxic materials from victims and personnel.

Emergency Decontamination

Another type of decontamination, emergency decontamination, can be conducted by personnel from first responding engine companies or truck companies with water supply capability (Figure 4.76). Emergency decontamination is a standard procedure for use with victims of exposure to terrorist incidents where chemical agents, radioactive materials, riot control agents, or biological agents have been deployed. Emergency

Figure 4.76 Emergency decontamination, can be conducted by personnel from first responding engine companies or truck companies with water supply capability.

decontamination could also be used for victims contaminated with hazardous materials to reduce the damage to the body from the chemicals. When there are contaminated victims or first responders become contaminated technical decon may not be readily available. Victims cannot always be allowed to wait to have contaminates on them until technical decon is set up depending on the type of material they have on them.

Emergency decon should be conducted as a defensive operation (Figure 4.77). Personnel conducting emergency decon, regardless of the circumstances, should be wearing full firefighter turnouts and SCBA. Water for emergency decon can be obtained from the booster tank of fire apparatus or through a supply line connected to a fire hydrant. Hand hose lines or master stream devices on wide-angle fog can be used to provide water for emergency decontamination. Nozzles can be placed on discharge ports of two engines side by side to provide a wide-angle fog for decontamination. The area between the engines can be used as a privacy protected corridor.

Personnel conducting emergency decon should avoid contact with victims, product, and runoff. While there are important environmental concerns for the runoff from any type of decontamination, life safety of victims should be the first priority when conducting emergency decon. The U.S. EPA has issued guidance for emergency decontamination of victims of terrorist agents, which reflects the priority of victims and then protect the environment as soon as possible after life safety has been addressed.

Emergency decontamination works well for hazmat incidents where someone has been splashed with a corrosive or other material that is absorbed through the skin. Damage to exposed skin can continue as long

Figure 4.77 Emergency decontamination should be conducted as a defensive operation.

as the material is on the person. Copious amounts of water should be applied to counter the effects and remove the material.

Emergency decontamination can be accomplished with equipment carried on apparatus for firefighting and rescue operations. Aerial master streams with adjustable nozzles can be put on wide-angle fog patterns and sprayed toward the ground, creating a decon shower. Commercial devices also are available specifically for emergency decontamination. Such devices provide a wide spray pattern for mass decontamination. They can be attached to a ladder pipe on an aerial apparatus or a discharge port on an engine or hooked directly into a hydrant to free apparatus for other functions.

Removing Clothing The first step in technical or emergency decontamination of victims is to remove their outer clothing. Most of the contamination will likely be on the clothing of victims. Once removed, clothing and other personnel belongings should be bagged and marked for later disposal upon consultation with health and environmental officials.

Removing outer clothing of victims, while reducing the hazard, raises the question of privacy. Victims may be children and/or adults. Provisions need to be made so that victims are segregated by age and gender.

Children will be apprehensive of the decon process and the removal of outer clothing – the younger they are, the more apprehensive they may be. In some cases, children may be allowed to participate in some stages of decon, such as rinsing one another with a hose. Letting children participate in the decon process can greatly reduce the chaos and make the situation seem fun while accomplishing the task at hand.

Privacy for victims can be accomplished in a number of ways. Most fire department companies carry tarps of various sizes. With a little preparation, procedures can be developed to use tarps in combination with ladders, fire apparatus, or natural features at the site to provide privacy screens. For example, the Seattle Fire Department has developed a procedure for setting up a privacy corridor using an aerial ladder extended straight out. Tarps are hooked to the ladder with prefabricated hooks. The base of the tarp is held in place with a hose line.

Commercially available tents and decon trailers can also be used to provide privacy. Once the decon process is complete, victims will need to be covered with some type of temporary clothing. Paper clothing is available, as are hospital gowns and scrubs. Plastic trash bags with holes cut for head and arms can also be useful. Tarps and plastic sheeting can be used to help control runoff from emergency decontamination. However, the number one priority is the decontamination of victims; if personnel and equipment are not available for controlling runoff, then it should not be a concern at that point. Victims should always take priority over property or the environment.

Weather can provide challenges for responders for both emergency and technical decontamination. Cold weather climates are a particular problem. Ice can form from runoff, creating a dangerous fall hazard for responders and victims. Cold climates also present an exposure problem for victims, particularly during emergency decontamination. If heated water is available, it should be used during cold weather, but responders may have to be creative. Decon tents and trailers are usually outfitted with water-heating capability. Pools and make-shift emergency decontamination lines usually do not have heated water available. Water-heating decon devices are available commercially independent of trailers or tents, but they can be expensive.

If you are in a cold climate, you need to make some provision for heating decontamination water. The Anchorage Fire Department in Alaska has developed a dry decontamination process for cold weather to use on personnel who have been exposed to vapors or gases. This process makes use of positive-pressure evacuation fans that blow vapors and gases off of protective equipment. Personnel are then assisted with undressing and protective clothing is bagged for evaluation or disposed of if the PPE is disposable. Decontamination can present many challenges, but they can all be overcome with proper planning, training, and equipment. Every fire department, large and small, should be able to conduct effective emergency decontamination with equipment already available (FEMA/NFA).

Preparing the Hospital for Hazmat Patients Fire departments, EMS providers, and other emergency response agencies are being trained to deal with terrorist incidents, but recent testimony before Congress reveals that local public health systems are not prepared to respond to potential chemical or biological attacks. Testimony came from state emergency management officials and medical personnel. Reasons given include a lack of coordination among medical, emergency management, and law enforcement agencies, and the inability to detect a biological attack in a timely manner. This preparedness shortfall has been acknowledged by the Clinton Administration, and it is reported that the Department of Health and Human Services is working to improve the situation.

Concerns about a lack of preparedness for chemical and biological incidents on the part of the medical system may, in some cases, go beyond the terrorism arena. Many hospital emergency departments are not even prepared to deal with victims of accidental hazardous materials incidents. (Because agents used in terrorist attacks are hazardous materials, no differentiation is made here when referring to them when discussing procedures unless they are unique to a particular material.) Patients could contaminate a hospital emergency department and close it to other types of emergencies until it can be decontaminated.

While this has been identified as a national problem, the solution rests largely with the states and, in particular, local communities. Even if federal and state resources are available to assist in a hazmat release, such help would be hours away at best and most patients from a disaster reach the hospital within 90 min. Thus, hospital emergency departments must be prepared for hazmat patients and deal with them using their own resources for an extended period (Figure 4.78).

Figure 4.78 Hospital emergency departments must be prepared for hazmat patients and deal with them using their own resources for an extended period.

Much of the attention at a hazmat or terrorist scene is focused on treating victims and mitigating the incident. Response personnel also need to take into consideration treatment beyond the scene, that is, the hospital, and in particular the emergency room. Many hospitals, particularly smaller ones, are not prepared to accept potentially contaminated victims from a hazardous materials or terrorist incident. While it is not necessary that every hospital emergency department has the capability of receiving contaminated victims, response personnel should know which hospitals do have that ability.

Three situations can arise from hazmat or terrorist incidents:

- One patient has been contaminated and is transported to a hospital emergency department.
- A single incident has occurred, causing many casualties.
- A disaster disrupts a large segment of a community.

The successful outcome of a hazmat or terrorist incident will depend on good preplanning, which must include the local medical system. In addition to decontamination ability, hospitals should be identified that have the expertise to handle poison exposures, including stocks of antidotes.

When a community has more than one hospital, certain ones may be designated as hazmat facilities for the purpose of receiving patients from the emergency medical system. All other hospitals should at least have the ability to provide basic decontamination and care for people exposed to hazardous materials. Contaminated victims may show up at hospital emergency rooms on their own. After the sarin nerve agent attack in the Tokyo subway, only 600–700 of the estimated 5,500 victims were transported to hospitals by ambulance. The remaining victims arrived in private cars, taxis, and city buses (*Firehouse Magazine*).

Mass Decontamination

Mass-casualty incidents involving hazardous materials may overwhelm hospitals in a community (Figure 4.79). There may not be enough doctors and nurses, supplies, equipment, or bed space to treat victims. Without a preplan that makes contingencies for mass-exposure incidents, the impact of the disaster on the community will only be compounded by the medical system. Once plans are developed, they must be practiced and personnel should be trained to follow them.

Hazmat and terrorist training generally focus on emergency responders. In many areas, however, little attention is given to training doctors, nurses, and other emergency room personnel to decontaminate victims and provide treatment for chemical and biological exposure.

Preparation to treat biological exposure is of critical importance. Emergency departments may be the first to see patients from biological

Figure 4.79 Mass-casualty incidents involving hazardous materials may overwhelm all hospitals in a community.

exposure. Victims may report to emergency departments, clinics, or doctors' offices. Some victims may even be brought in by ambulance without EMS personnel realizing that the patients are victims of biological exposure. Symptoms of biological agents can be confused with the flu or other illnesses, and patients may be sent home.

When dealing with a known biological incident, decontamination should be performed. If the presence of a biological agent is not confirmed, as in the case of the recent anthrax scares, soap-and-water decontamination is sufficient. The CDC may be consulted to determine proper disinfectant solutions (FEMA/NFA).

Preparing the Emergency Department Emergency room personnel should be able to recognize hazards from chemical and biological exposure and determine appropriate protective equipment. When reports from the field of multiple patient conditions include nausea, dizziness, itching/burning eyes or skin, or cyanosis, personnel should recognize that hazardous materials could be involved. It is important that emergency room personnel have positive identification of the hazardous material and determine its toxic characteristics.

If the decision is made to maintain an ability to perform decontamination in house, appropriate PPE and training for donning and use will be necessary. Personnel must be trained in the use of PPE before they use it in an incident. Training and equipment must also conform to OSHA and other applicable regulations and standards. OSHA's 29 CFR1910.120 final rule as it applies to emergency medical personnel states that: "Training shall be based on the duties and functions to be performed by each responder of an emergency response organization." In addition, protective equipment in emergency departments also requires routine maintenance, cleaning, and resupply, when used.

Patient Management A fire department, hazmat team, EMS crew, or other responding agency should notify an emergency department that a patient or patients exposed to hazardous materials are en route (Figure 4.80). This should set in motion a preplanned course of action. Such a preplan calls for personnel to be aware of their responsibilities and how to perform them, and necessary equipment to be readily available or easily accessed.

- Type and nature of the incident.
- Caller's telephone number.
- Number of patients.
- Signs/symptoms being experienced by patients.
- Nature of injuries.
- Name of chemical(s) involved.
- Extent of patient decontamination in the field.
- Estimated time of arrival.

EMS personnel should be notified if the patient is to be brought to a special location or entrance to the emergency department to control potential contamination or to perform decontamination. Upon notification of a

Figure 4.80 A fire department, hazmat team, EMS crew, or other responding agency will notify an emergency department that a patient or patients exposed to hazardous materials are en route. (Courtesy: Vanderbilt University Hospital.)

contaminated patient or patients en route, notifications are made according to the preplan, the decontamination area prepared, and the decon team suited up.

Decontamination rooms should be well ventilated and, if possible, have ventilation systems that are separate from the rest of the hospital. Drains from the decontamination rooms should go to segregated holding tanks. These rooms should not contain sensitive equipment or supplies which could become contaminated and have to be trashed. Prevention of contamination to the room should include protecting all doorknobs, cabinet handles, light switches, and other areas where contamination may spread. Floors should also be covered with plastic and taped to prevent slipping.

A basic decontamination setup should contain the following:

- Location for patients to undergo decontamination.
- Means to wash contaminants from patients.
- Containment for the runoff.
- Protection for personnel handling patients.
- Medical equipment to treat patients that is washable or disposable.

Decontamination team members should be predesignated and trained in appropriate protective equipment and procedures. Team members should include emergency physicians, emergency department nurses and aides, and support personnel such as security officers, maintenance workers, and biological safety officers. When patients arrive at the hospital, they should be met by the emergency physician-in-charge or an emergency department nurse to assess their condition and degree of contamination. Triage procedures may be implemented if needed. All contaminated clothing should be removed. Clothing will need to be double-bagged in plastic, sealed, and labeled.

If decontamination is necessary, the decon team should bring a prepared stretcher to the ambulance and transport the patient directly to the decon area. Open wounds should be protected to prevent them from being contaminated. Priority is always given to the ABCs (airway, breathing, and circulation) while personnel conduct decontamination. Effective decontamination involves thorough washing of patients. The contaminant should be reduced to a level that is no longer a threat to patients or personnel.

Because victims may arrive at an emergency room on their own, the hospital must have the ability to conduct decontamination when needed and protect personnel before treatment can begin. Decontamination in the hospital setting can range from a typical decontamination line set up outside the emergency room or in the parking lot, to a dedicated decontamination room inside the hospital. In some cases, local hazmat teams may be called upon to set up decontamination outside emergency rooms (*Firehouse Magazine*).

Prepare for the Worst Differences exist between a mass-casualty disaster and one involving hazardous materials (Figure 4.81). These include the need for decontamination of patients and personnel, and identifying effective safety measures to protect personnel. Standard triage procedures are used for hazmat disasters, except in cases of extremely toxic materials. Without antidotes, many victims of toxic exposure may not be treatable and efforts should be focused on those who can be saved.

Without trained and equipped emergency responders and hospital emergency departments, hazmat incidents and biological terrorist attacks may lead to a disaster for the local medical system. Proper planning, training, and equipment at the emergency department will greatly reduce the impact of an incident on the community.

Decontamination RIT for Hazmat Personnel Think of decontamination as the rapid intervention team (RIT) for hazardous materials personnel. Generally, we do not send in an RIT team unless a "May Day" has been called. We do not always have to deploy a decon team for every incident we respond to. Even if we do decide to set up decon, those making entry should practice contamination avoidance. Remember, to SLOW DOWN when responding to and planning action options for the circumstances of the hazardous materials container escape you are facing. Until you have identified the hazardous material, researched its hazards, conducted a

Figure 4.81 Differences exist between a mass-casualty disaster and one involving hazardous materials. (Courtesy: Seattle Fire Department.)

Figure 4.82 Many jurisdictions do not have their own hazmat teams. They require the use of mutual aid or private contractors for hazmat responses.

risk analysis, and determined an entry will be made, decon setup can wait. Today there are multiple decontamination options available other than soap and water.

What type of decon will be needed won't be known until the other steps just mentioned have been practiced, so SLOW DOWN. You might not need any decon at all. Think of all of your options on the scene of a hazardous materials escape as tools in your Hazmat Tool Box. When you are working on repair projects at the fire station or at home, do you bring every tool you have to the repair do the repair job site? I do not. You have your toolbox on-site at a hazardous materials scene. You only take out what you need to do the job based on the hazard of the material or materials that are present. SLOW DOWN, you have time to determine what tools are needed, and when they need to be taken out of the toolbox. Just like home projects, the more tools you pull out, the more you have to put back. That takes time from other tasks or in the case of the fire department, other emergencies.

Dry Decontamination Wipes

> ***Author's Note:*** *As I have traveled around the country visiting hazmat teams, I have noticed a change toward the use of dry decontamination wipes for decon. Dry decon wipes have been developed by the FBI Hazmat Response Unit and have been proven to be an effective decontamination method without the challenges created by traditional methods. Wipes have also been developed by Edgewood Chemical and Biological Command (ECBC) for the immediate decontamination of chemical warfare agents.*

Wipes work well on nonporous surfaces, much the same as traditional decontamination. Once a contaminant permeates a porous surface it is unlikely that the manufacturer of the porous material will sign off on the decontamination of their product. This is largely because there are no nondestructive tests that can be done to determine total decontamination and the liability of the manufacturer if they make a judgment that proves to be wrong. Responders should always practice contamination avoidance at all times. This includes both PPE and equipment. ECBC has reported some success with adsorbents in removing contamination from some equipment. However, some of the products mentioned are still in the development and testing stages. Products are available commercially from several vendors. This is still an evolving concept in decontamination. Response organizations need to evaluate their decontamination needs and products available to see if they fit into their organization's operations.

"Dry Decontamination" Fans Anchorage firefighters have developed a method of decontaminating personnel during extremely cold conditions, using "dry decontamination" procedures (Figure 4.83). These procedures are used when personnel are exposed to a gaseous atmosphere or are not significantly contaminated by a product. Positive-pressure ventilation (PPV) fans are set up to blow off any residual gas or vapor that may be pocketed anywhere on the suit. Personnel then step into a large plastic bag that has been prepositioned on a tarp and are assisted by other team members to remove their suits. The person within the suit does not

Figure 4.83 Anchorage firefighters have developed a method of decontaminating personnel during extremely cold conditions, using "dry decontamination" procedures.

participate in the removal. Other team members roll the suit with the contaminated side away from the wearer. Once the suit is rolled all the way down, the wearer steps out of the suit. The suit is then packaged within the plastic bag and properly disposed of.

Decon Foam "Researchers at the Department of Energy's Sandia National Laboratories have created a type of foam that begins neutralizing both chemical and biological agents in minutes (Figure 4.84). Because it is not harmful to people, it can be dispensed on the incident scene immediately, even before casualties are evacuated. The foam, comprised of a cocktail of ordinary substances found in common household products, neutralizes chemical agents in much the same way a detergent lifts away an oily spot from a stained shirt. Its surfactants (like those in hair conditioner) and mild oxidizing substances (like those found in toothpaste) begin to chemically digest the chemical agent, seeking out the phosphate or sulfide bonds holding the molecules together and chopping the molecules into nontoxic pieces. How the foam kills spores (bacteria in a rugged, dormant state) still is not well understood. The researchers suspect the surfactants poke holes in the spore's protein armor, allowing the oxidizing agents to attack the genetic material inside." (Sandia National Laboratories, 1999)

Figure 4.84 Researchers at the Department of Energy's Sandia National Laboratories have created a type of foam that begins neutralizing both chemical and biological agents in minutes.

Military Agent Decontamination

Unlike most hazardous materials incidents, decontamination for chemical agent releases will focus primarily on contaminated victims. There are likely to be hundreds and even thousands of exposed citizens who may need various levels of decontamination. The process will be very labor-intensive and unlike any other decontamination process responders have previously faced. In some cases, citizens will need to be decontaminated simply for their own peace of mind. Vapor exposures do not require as extensive decontamination as does liquid exposure.

In reality, little contamination occurs from vapor exposure. The public, however, may perceive a need for decontamination, and it may have to be performed, if only, for psychological reasons. Decontaminating hundreds or thousands of potentially contaminated victims could create a logistical nightmare. Responders in the CSEPP program have acquired specially designed decontamination trailers for treating large numbers of citizens. These trailers would work well for mass exposures from terrorist incidents. When large numbers of casualties start showing up from apparently unknown sources, chemical or biological terrorism should be a consideration. These types of incident swill tax the emergency response system as never before. Responders must plan and train for chemical as well as nuclear and biological terrorist attacks but even with planning, these will be difficult incidents.

Robots

Pentagon Force Protection Team

One of the many assets maintained by the CBRNE division is a Remotec MK3 Mod 0 robot provided by the Hazardous Devices Division. The robot was outfitted with monitoring equipment for chemical warfare agents, toxic industrial chemicals/toxic industrial materials (TICs/TIMs), and radiological materials (Figure 4.85). The robot will be used to survey an incident prior to team members making an entry. This will allow the gathering of basic atmospheric readings, enhance incident size-up, and assist in the selection of PPE. The robot will also be used to provide video and audio from the "hot zone" to the command post. The robot can also be used to haul equipment into the "hot zone," reducing the workload of responders wearing PPE. Pentagon Force Protection Team has access to dozens of robots from the department of defense that are no longer in combat. Some are used for parts, but others can be fixed up and used for hazmat (*Firehouse Magazine*).

Gwinnett County, GA Gwinnett County Hazardous Device Unit carries several remotely operated vehicles for use at hazardous materials as well

Figure 4.85 The robot was outfitted with monitoring equipment for chemical warfare agents, toxic industrial chemicals/toxic industrial materials (TICs/TIMs), and radiological materials.

Figure 4.86 Gwinnett County Hazardous Device Unit carries several remotely operated vehicles for use at hazardous materials as well as hazardous device scenes.

as hazardous device scenes (Figure 4.86). These vehicles allow for a preliminary evaluation remotely without having to send operators or other personnel downrange into harm's way. In addition to monitoring instruments carried on the vehicles, they can deploy X-ray equipment to X-ray suspicious or actual explosive devices and bring the film back to personnel for development and viewing. The X-ray can also be utilized to look at pipes or valves on hazardous materials containers that may be wrapped in insulation and leaking.

Video equipment on the vehicles can be used to view hazardous materials containers for markings and other identification information about the hazardous materials without personnel being exposed to danger. Because the hazardous materials team members are familiar with much of the hazardous devices unit equipment, if additional manpower is needed at an incident scene, they can be requested to respond. Training is also conducted with hazardous materials team members to provide EMT and paramedic support. EMTs and paramedics are taught how to treat police personnel who might become injured during a hazardous device response. In particular, they are instructed on how to remove bomb suits during a medical emergency at an incident scene (*Firehouse Magazine*).

Orlando, FL Orlando uniquely operates its bomb squad out of the fire department. Robots can also benefit hazardous materials response (Figure 4.87). A robot is available for situations where it is too dangerous to risk personnel for tactical operations. The robot can move around on rubber wheels or be quickly changed to tracks for difficult terrain. Service is available 24 h on-site if the robot breaks down; if it can't be quickly fixed, the company will fly in a replacement. A video camera on the robot has the capability to feed video back to the bomb truck from the scene. It is also equipped with lights and two-way communications. A video camera is located on the roof of the bomb truck to provide feeds inside from the surrounding area. None of the typical hazmat equipment

Figure 4.87 In Orlando, a robot is available for situations where it is too dangerous to risk personnel for tactical operations for hazmat and law enforcement.

Figure 4.88 None of the typical hazmat equipment is carried on the bomb truck in Orlando, it is all on Hazmat 1.

is carried on the bomb truck it is all on Hazmat 1 (Figure 4.88) (*Firehouse Magazine*).

Picatinny Arsenal, NJ Picatinny's Hazmat Team operates a Talon robot, named "FRED" by the "B" Platoon (Figure 4.89). The name is short for **F**ind and **R**eference **E**nvironmental **D**ata. The robot folds up and is carried compactly in a front compartment on the driver's side of the hazardous materials unit so it is readily available on all responses. It is battery operated and charged constantly while in its storage compartment. FRED is

Figure 4.89 Picatinny's Hazmat Team has operated a Talon robot, named "FRED" by the "B" Platoon.

equipped with a Raytek temperature sensor which can probe the surface of a container to check for temperature change. It also contains monitoring instruments including an APD 2000 for detecting military chemical agents. A Drager Multiwarn monitor is used for detecting anhydrous ammonia, chlorine, hydrogen cyanide, methane, and oxygen concentrations. For radiation detection, a Canberra Rad detector monitors for alpha, beta, and gamma radiation. All monitors are capable of functioning simultaneously. FRED is also equipped with four video cameras, an elbow camera, a gripper camera, a rear camera, and a 360° camera. The gripper camera is located on the end of the robotic arm called the gripper and allows the operator to see what the gripper is doing while controlling it remotely.

FRED's gripper has a grip strength of 35 psi. This allows the robot to lift 25–30 lb and drag 200 lb depending on the surface. The lifting feature allows for the movement of materials on scene or retrieval of items or information from the scene. The robot is also capable of a dragging function. This function can be used by the robot to remove victims including response personnel or to ferry equipment into or out of the scene (Figure 4.90). Two-way radio communications located on the robot allow for monitoring of sounds in the hot zone or initiating contact with contaminated or injured victims.

When a bomb is suspected the robot has a spool of fiberoptics and can establish communications without using radios that might set off a bomb. FRED is normally outfitted with a universal track but has others for varied types of surfaces. A hazmat tech controls the operation of FRED from the Operations Control Unit (OCU). This is basically a computerized unit that can display the video from the robot's cameras, individually or

Figure 4.90 The robot is also capable of a dragging function.

all four at the same time in a split-screen view. The unit provides real-time data readouts from the robot's monitors. All movements of the robot including the gripper are controlled from the OCU along with communications capabilities as well. FRED can operate from 6 to 8h on a single battery charge.

The robot is often used to gather information on the hazardous materials involved in an incident remotely, thus keeping personnel out of harm's way until more information can be determined on the hazards of the materials. Once entering the hot zone the robot can send back video showing placards or other markings on containers, the leak in the container, or the spill itself. Other information can be obtained from the video including any other circumstances that may compound the problem with the hazardous materials. Samples of product spilled could also be obtained by the robot expediting the identification process. The robot can be removed from its storage compartment on the hazmat unit by two team members and deployed in a matter of minutes. Setting up decon and dressing-out entry personnel to gather information and then decontaminating them would take much longer (*Firehouse Magazine*).

Philadelphia Hazmat Robot (Figure 4.91)

Firefighting Robots for Fires Involving Hazardous Materials

According to the NFPA, there were 29,130 injuries reported while fighting fire in 2015 (Figure 4.92). These injuries are also coupled with 68 on-duty deaths. Injuries and casualties are the reason why safety professionals, the

Figure 4.91 Philadelphia Hazmat Robot.

Figure 4.92 According to the NFPA, 29,130 injuries were reported while fighting fire in 2015.

government, and high-tech companies have come together to create firefighting robots that can perform tasks too risky for people (Philadelphia Fire Department).

Firefighting Technology Mobile firefighting robots are another type of robot that could be useful in fighting fires involving hazardous materials and reducing firefighter deaths and injuries. These robots are remote-controlled vehicles affixed with fire suppression tools like water or foam hoses. These are capable of traveling into areas unsafe for people through an array of sensors, visual camera, infra-red, and other technology that transmit information for navigation to a remote operator. The following are some real-life advancements in firefighting robots that can reduce the number of injuries and fatalities caused by fire.

THOR/SAFFiR

The Tactical Hazardous Operations Robot (THOR) was developed for the U.S. Navy's Shipboard Autonomous Firefighting Robot (SAFFiR) program. Its a humanoid robot with the capability of traversing unstable floors on ships as well as utilizing hoses and opening doors. Navy ships have hazardous materials onboard with minimal room to move, so extinguishing fires is paramount. It stands at 177cm tall, uses stereoscopic thermal imaging and LIDAR (light detection and ranging) sensors for navigation, and has successfully put out compartment fires along with the assistance of a person. The goal is for THOR to walk and work semi-autonomously with the help of a remote operator. Currently, however, there are still some setbacks. The machine is slow and susceptible to fire and water damage. These challenges are being resolved so that it may extinguish fires too dangerous for people to get near (THOR/SAFFiR).

Thermite Robot

Originally a small tank created for the U.S. Army by Howe and Howe Technologies, this remote-controlled vehicle has been suited with a hose that is capable of pumping 500 gallons of water per minute. Using mounted cameras, it can travel into extremely hazardous situations, like wildfires, while being controlled from up to a quarter-mile away.

Weighing in at 1,640 lb, it was designed as an improvised explosive devices neutralizer. With some firefighting modifications, it can also be used as a fire neutralizer. Onboard is an innovative cooling system that provides consistency by using some of the water pumped as a coolant throughout its body. Although it pumps one-third less water than a fire engine (1,500 gallons/min) it also costs considerably less. Coupled with its ability to enter hazardous areas without putting people in harm, it could be seen putting out wildfires in the future (Thermite).

Turbine Aided Firefighting Machine (TAF 20)

Created by Emicontrols, a subsidiary of the TechnoAlpin Group, TAF 20 uses a turbine as an innovative method in firefighting. Meant for small spaces like tunnels, it has the ability to move obstacles with its bulldozer blade, clear smoke with its turbine, and focus its water spray from mist to jet. Its a tracked vehicle that uses the turbine to atomize water into a large mist that covers more area while using less water. In certain cases, the water can be focused into a powerful jet capable of spraying 3,500 L/min. Operators can be up to 500 m away, but it is restricted like other firefighting robots due to its connection with a water hose (Turbine).

Fire Ox

The Fire Ox is one of the few robotic firefighting vehicles that carries its own water tank. Designed as a first response unit, it suppresses fires, assists in search and rescue, and can handle dangerous materials. Originally created by Lockheed Martin as a Squad Mission Support System (SMSS) for assisting soldiers with their gear in the field, it was retrofitted with a water tank and hose for distribution. The Fire Ox is semi-autonomous and can be controlled from up to 200 miles away. Due to its mobility, it has the ability to traverse situations unsafe for people, minimizing casualties and rescue time. It can also be used in a multitude of situations including wildfires and structure fires (Fire ox).

Future Development of Robotics in Firefighting
Firefighting robots are still being developed and perfected for more common use in the field. Occupational safety professionals are working with other professionals and governmental agencies to make futuristic technology a reality. Those who have a passion for defending first responders from overly dangerous situations can help shape the future of firefighting robotics.

The safety of firefighters and the victims caught in fires is paramount, urging the production of these robots forward. They have the ability to withstand environments too hazardous for people and can prevent further damage that may happen left to traditional methods of suppression.

Drones

Greater Cincinnati, OH Hazmat Team

GCHMU is constantly looking to improve response operations and finding methods to make response personnel safer. Several technician-level tactics that require entry into a hot zone place them in harm's way. Chief Bennett and Duty Officer B.J. Jetter have been looking into drones for application to hazmat response, with help from the University of Cincinnati. While they do not yet have on-scene experience with drones, Chief Bennett is a licensed drone pilot and owns his own Phantom 4. Fire department and hazmat use drove his interest in purchasing a drone. He is concerned about sending personnel into the hot zone and would prefer to utilize a drone instead.

While the FAA has a 400 ft height limit for drones, Chief Bennett says "you would be surprised at what you can see at 400 ft." Certainly, a drone can provide a better view and send video and photographs that can be reviewed at the command post over and over to see all of the details. An added value is that a drone can also zoom in as close as required for a clear view. Using a drone places the drone in harm's way, but that is better than sending personnel in when the drone can accomplish the same thing. Chief Bennett also prefers to purchase a cheaper drone, if it will do the job, so if the drone is lost, it will cost less to replace it.

Train derailments are one type of incident that drones are particularly suited for. Tracks do not always follow highways and may occur in areas where access is limited. Sending in a drone would provide valuable information in an area that accessing by foot in a chemical suit or even bunker gear would be difficult at best. With the frequency of ethanol and crude oil shipments, trains with multiple tank cars of the same material are not uncommon. If multiple cars derail, using a drone to size-up the situation would be invaluable.

Chief Jetter also wants to use a drone for air monitoring and sampling. This would require a second drone as most drones do not allow for multiple functions at the same time. Also, they do not allow for interchange of tools such as cameras and air monitoring equipment. This was Chief Bennett's concern as well. Being able to locate a leak at an incident scene and then monitor the air and or take a sample would not only reduce time on scene but keep responders out of danger. Because most departments do not do clean up, with a drone, in some circumstances, entry by personnel might not be necessary.

There is also the added problem of decontamination of the drone, which might not always be possible. However, if the drone is in a vapor, not much contamination will occur from vapors. On the other hand, vapor may damage the drone. Drones are not the replacement for robots or even hazmat personnel they are just another tool in the toolbox. However, having drones with interchangeability of equipment would certainly make them more valuable. Once drones are purchased, personnel need to be trained and certified to use them.

The University of Cincinnati has provided seminars and credit courses for non-degree seeking students. One such course is *Unmanned Aerial Vehicles [Drones] for Emergency Responders.* FAA has regulations for the ownership and use of drones (107 Certification). Drones are limited to flight 400 ft above the ground and can only be flown in the sight of the operator. Exemptions have been granted, however getting one is not an easy task. Operating the drone requires additional insurance. According to Chief Bennett, "the industry has not caught up with the technology." Many carriers do not insure drones. Amounts of coverage that will be issued have increased from 1 to 2.5 million. Some insurers will provide insurance by the hour as low as $10.00/h (*Firehouse Magazine*).

Corpus Christi, TX

Corpus Christi Fire Department has established an AERO Team (Figure 4.93).

I visited Jim in Corpus Christi to write about the progress they had made in their drone program. Jim is now the AERO Team Coordinator for the Corpus Christi Fire Department. Battalion Chief J.D. Johnson is now the Coordinator for the Hazmat Team. AERO stands for Aerial and Emerging Robotics. Corpus Christi's drone program is off and running. Jim had been sending me drone photos and video from time to time from hazmat incidents, so I knew they were up and running (Figure 4.94). Any incident commander or outside agency can request the AERO team through Fire Chief Robert Rocha or AERO Team Coordinator Jim DeVisser. Team drone operators and Chief DeVisser respond to the incident with the drones. Jim acts as the mission coordinator and interacts with the requesting IC or outside agency (Corpus Christi Fire Department).

Figure 4.93 Corpus Christi Fire Department has established an AERO team.

Figure 4.94 Drone view of overturned tanker truck burning on expressway off ramp. (Courtesy: Corpus Christi Fire Department.)

Special Intervention Procedure Allegheny County, Pennsylvania

Hazmat Rapid Intervention Team

Special Intervention Procedure Allegheny County, Pennsylvania Green Team One of the unique specialties of the Allegheny County Pennsylvania Green Team is their Rapid Intervention Procedure (RIT)

Figure 4.95 West County Fire Department Saint Louis County, MO has had a drone program for at least a decade.

Figure 4.96 Hazmat RIT developed by team members for the rescue of hazmat team personnel who might become trapped from falling debris or other cause during entry into the hot zone.

that was developed by team members for the rescue of hazmat team personnel who might become trapped from falling debris or other cause during entry into the hot zone (Figure 4.96). The RIT procedure was developed by Deputy Chief Ryan Lattner – Allegheny County Airport Authority, Deputy Chief Kurt Gardner – Allegheny County Specialized

Intervention Team and Commander Jim Eaborn – Allegheny County Specialized Intervention Team. The primary purpose of the RIT is to provide an emergency air supply for personnel in Level "A" protective clothing who become disabled during an entry until they can be extricated, decontaminated, and receive medical attention as required by any injuries. The emergency air supply procedure is a last resort and is only used in situations where death or serious injury would occur from lack of air. The RIT is established and on stand-by anytime a Level A entry is performed. They are partially suited up and ready for rapid deployment. Special kits are assembled ahead of time and contain a modified SCBA, cutting tools and Chem Tape. Strips of tape are precut and placed on the SCBA tank for quick access. When the team makes an entry they bring the kit and a skid for removal of the team member when extricated if needed. When a responder goes down the RIT team enters the site and locates the team member. They determine if the scene is safe for them to enter and if a rapid extrication is possible. If rapid extrication is not possible they assess the level of consciousness of the responder that is trapped and their available air supply. When the air supply reaches a critical level the RIT team will breach the suit of the responder and supply additional air. A cut approximately 2–3 in. is made in the Level "A" suit (Figure 4.97). The hose from the rescue SCBA is inserted into the Level "A" suit through the cut. Precut pieces of Chem Tape are used to secure the hose in place and reduce air leakage through the hole. Once the hose is secured the air supply is turned on slightly so that the suit is not inflated. If the trapped responder is conscious they are instructed to remove their arms from the sleeves of the Level "A" suit and remove

Figure 4.97 A cut approximately 2–3 in. is made in the Level "A" suit.

their SCBA mask. If the responder is not conscious RIT team members remove the mask from the outside of the suit. RIT team members monitor the air supply of the trapped responder and assist with extrication. Once extricated the responder is then removed from the entry site and decontaminated.

When first contacted by the Allegheny Team about the RIT procedure I had some reservations. My feeling was situations where a person runs out of air should be totally avoided by following proper procedures and safety precautions. However, after gaining further knowledge about the intention and application of the procedure and witnessing it demonstrated first hand I was extremely impressed. While gathering information for this column I personally witnessed a drill utilizing the RIT procedure at the National Energy Technical Labs in Allegheny County. The procedure was quickly and effectively deployed for a downed person in Level A protective clothing trapped under heavy debris that had fallen. The Green Team is well trained and equipped to carry out this rescue procedure and they should be applauded for their ingenuity for the creation of this valuable rescue tool (*Firehouse Magazine*).

Basic Chemical Storage Segregation

There are likely to be few places where firefighters and other emergency response personnel will encounter a wider variety of dangerous chemicals than in a chemical laboratory. Laboratories can be found in a variety of locations, including industrial, research facilities, high school, and college labs to name a few. High school and college laboratories are particularly dangerous because the variety of chemicals used and stored there may be greater than any other lab. While the quantities of individual chemicals are usually relatively small, collectively, and in some cases individually, the danger to response personnel can be significant. Chemicals can be toxic in very small amounts when absorbed through the skin, inhaled, or ingested. They can also cause eye damage, skin burns, illness, and cancer.

Every category of the Department of Transportation Hazard Class can be identified among the chemicals in a high school or college laboratory setting. Other labs are more specialized and the range of chemicals is limited to research projects or analytical needs of the facility. When stored with caution, most chemicals do not pose an unreasonable threat under normal conditions. However, chemicals in laboratories are often stored in improper locations and alphabetical order, which may place completely incompatible chemicals on the same shelf with each other. Storing chemicals alphabetically can place dangerous materials such as nitric acid, which is a strong oxidizer, on the same shelf with flammable liquids.

Mixing the two can result in an explosive compound. A more appropriate storage system might be one in which organic and inorganic chemical families are stored together. Many chemical supply companies include proper storage system information in their chemical catalogs. Fisher Scientific company has a very good one in their catalog. Material Safety Data Sheets (MSDS) may also contain information about chemical compatibilities. MSDS should be maintained on file for each chemical in the laboratory. Manufacturers or chemical suppliers can provide MSDS upon request. In many cases, they will be shipped automatically with the chemicals. Chemicals that require refrigeration and storage in flammable liquids or acid cabinets are often stored on open shelves in the lab or classroom. Acids and bases, while categorized by the Department of Transportation as corrosives, react violently with each other and should be stored in separate locations and cabinets. Other chemicals can be dangerous under normal conditions when exposed to heat, shock, friction, water, or air.

Care must be taken to store these chemicals in safe locations. To make ordinary storage conditions worse, chemicals can degrade, dehydrate, or form other dangerous compounds as they age. Many compounds that are normally safe can be converted into shock- or heat-sensitive explosive materials as they age. These factors can create risk for emergency response personnel if they are unaware of the dangers. Oftentimes laboratory personnel and teachers are unfamiliar with the hazards of aging chemicals. Fire inspectors need to be aware of these dangers so they can point them out during laboratory inspections.

In many industrial and educational occupancies, laboratory operations take up only a small portion of the space in the facility. Plating labs are used to maintain the quality of plating solutions within the plant. Chemicals found in these labs are fairly limited and include cyanide, sodium hydroxide, and many different types of acids. Contact between acids and cyanide can produce deadly hydrogen cyanide gas. Other types of industrial labs are used to maintain quality control of manufactured products, and chemicals may vary from one facility to the next. Preplanning is an important step in determining the hazards of these facilities.

National Fire Protection Association (NFPA) Standard 45, as well as NFPA 30, 49, 325, 491M, and 704, are good sources of information concerning fire code issues in laboratory occupancies. Research facilities located in industrial plants or universities also use limited but very specific chemicals based upon their research. Some of these facilities also have an additional biological hazard over and above the chemical hazards present. Once again, preplanning should ease the minds of response personnel when fires or other emergencies occur at these locations. Marking systems such as NFPA 704 or Hazard Communication should be used to identify areas within buildings where dangerous chemical or biological materials may be used and stored along with emergency contact information for use by response personnel during an emergency.

Firefighting operations usually involve the application of water through a hose line. Small fires in laboratories can be extinguished safely by using portable dry chemical fire extinguishers. When metals such as potassium are involved, Class D dry powder fire extinguishers should be used. Any other type of extinguisher will not be effective. Inserting a hose line into a laboratory chemical storage area can cause glass containers to break and mix chemicals together. Even a chemist couldn't tell you what the potential outcome would be if chemicals were mixed that are not normally placed together. Firefighters and other rescue personnel may encounter highly toxic and carcinogenic chemicals and mixtures along with flammable, water and air reactive, and explosive materials. Great care should be taken when fires occur in laboratories. Runoff from firefighting can be very toxic and can cause environmental damage, or at the very least pose danger to personnel and contamination of personal protective equipment. Some fires may be better left to the sprinkler system to extinguish or allowed to burn themselves out.

To ensure the safety of personnel, some important steps should be taken in handling chemical fires and emergencies.

- Firefighters and other rescue personnel should never approach a potential chemical emergency scene or fire without self-contained breathing apparatus and proper protective clothing. Care should be taken not to contact any chemicals or runoff with turnouts.
- During the overhaul, extreme care should be taken to ensure firefighters are not exposed to chemicals. Runoff water, which may have become contaminated during firefighting operations, should also be controlled.

Prevention is always crucial. Some effective preventive measures include:

- Preplanning schools, colleges, universities, and other locations where laboratories are located. Note locations of hazardous chemicals and use the NFPA 704 or another type of marking system to identify the locations of chemical use and storage.
- Conduct regular fire inspections of these facilities.
- Obtain inventories of chemicals stored in the laboratory areas and make them available to emergency response personnel during emergencies.
- Remove explosive, reactive, severely flammable, toxic or no-longer-used chemicals from high school and college laboratories through the use of qualified chemical and explosives experts.

- Encourage teachers, researchers, and laboratory technicians to order small amounts of chemicals instead of large supplies that present a danger to response personnel or may become dangerous through long-term storage.
- Instruct laboratory workers and instructors to date substances when purchased and again when opened that may become dangerous with age. Unused portions should be properly disposed of within 6 months to 1 year.
- Encourage lab managers to set up proper storage systems so the chance of incompatible chemicals combining is reduced. Flammables should be stored in approved cabinets or 1-hour fire-rated storage rooms with proper ventilation and sprinkler protection. Acids should be stored in approved acid cabinets away from flammable liquids and solids. Nitric acid should always be stored away from other acids. Bases should be stored separately from acids.
- Provide familiarization tours for fire and other emergency response companies so that personnel will have a better understanding of the hazards in laboratories.

In addition to removing dangerous chemicals from laboratory settings, it is important to educate teachers, laboratory workers, and research personnel on the proper storage, use, and purchasing practices through training classes and literature. High school and college instructors should also be given assistance to develop experiments using safer chemicals that produce similar results but reduce the danger to students, teachers, and emergency responders. Through the cooperative efforts of school administrations, industry leaders, research organizations, and emergency response agencies, schools and laboratory settings can be made safer for occupants and response personnel. With proper training and preparation, emergencies involving hazardous chemicals in schools and laboratories can be safely and successfully handled (*Firehouse Magazine*).

Hazmat Response in Rural Areas

Most of the recent firefighter deaths in the United States involving hazardous materials have occurred in the nation's heartland. Incidents in rural places like Burnside, IL (population 242), Albert City, IA (population 709), Ghent, WV (unincorporated), and Tilford, SD (unincorporated) have resulted in the deaths of firefighters and other responders. Two firefighters died in Burnside and two in Albert City. In Ghent, a firefighter, EMT, and building inspector died along with a civilian. In Tilford, an assistant chief died.

Over the years hazmat incidents resulting in large numbers of firefighters and civilians being killed and injured have also occurred in rural areas (Figure 4.98). In Waverly, TN (population 4,028; less in 1978), 16 people were killed including the Waverly Fire Chief, Police Chief, and 5 firefighters. In Kingman, AZ (population 27,000; less in 1973), 11 firefighters and 1 civilian died. Other hazmat incidents have also occurred in rural America that resulted in catastrophic property loss but did not cause loss of life because of the prudent actions of emergency response personnel and perhaps a bit of luck.

Crescent City, IL (population 631) had a major train derailment in 1970 with several BLEVE's which resulted in injuries but no deaths. An Illinois State Trooper who responded to the incident and was familiar with propane and its hazards moved firefighters out of harm's way prior to the rail tank cars exploding, which likely saved their lives. Weyauwega, WI (population 1,806) had a major train derailment and fire in 1996, but no one was killed or injured and the entire community was evacuated for 21 days. This incident was similar to Waverly, TN, and the fire chief of Weyauwega told me it was the lessons learned from the Waverly, incident that caused them to take the precautions they did including the evacuation of the entire community and even a nursing home.

Ironically, all of the incidents mentioned involved the same type of hazardous materials: flammable liquefied petroleum (hydrocarbon) gases. These gases included propane, butane, and liquefied petroleum gas (LPG),

Figure 4.98 Over the years hazmat incidents resulting in large numbers of firefighters and civilians being killed and injured have also occurred in rural areas. (Courtesy: Crescent City, IL Fire Department.)

which is a mixture of hydrocarbon gases. All of the gases mentioned are extremely flammable, have large expansion (from liquid to gas) ratios, and are asphyxiates (displace oxygen in the air). They have low flashpoints, low boiling points, and are heavier than air. These gases are not refrigerated so the temperatures of the liquids are close to the ambient temperatures where they exist. Therefore, these materials exist as a liquid in their container well above their boiling point in most ambient conditions in the United States.

This is where the term BLEVE originates from. When a container is breached during a BLEVE, all of the liquid in the container is released and almost instantly expands and turns into a gas. This is a violent process that usually rips containers apart and rockets parts of the container thousands of feet from the initial location. If the material in the container is flammable and finds an ignition source a fireball may also result. It is this BLEVE that resulted in the deaths and injuries among communities previously mentioned. Several factors involving liquefied gas containers can result in a BLEVE. Flame impingement on the vapor space of the container weakens the metal rather quickly. National Fire Protection Association (NFPA) statistics indicate that BLEVEs occur between 8 and 30 minutes of the start of flame impingement, with 15 minutes being the average. Fifty-eight percent of BLEVEs have occurred in 15 minutes or less. Containers can also be damaged during derailments and other accidents, weakening the integrity of the container. Increases in ambient temperature or heat from adjacent fires can cause an increase in pressure within the container. At some point the weakened portion of the container can no longer handle the pressure increase and the container comes apart violently.

Rural America is often protected by small career, combination, or entirely volunteer fire departments. The U.S. Fire Administration reports that 87% of the fire departments in the United States are volunteer or mostly volunteer and protect 38% of the population. Many of these communities are small by population but have large hazardous materials exposure. Some may have rail lines that transport hazardous materials through their community while some may not. Others may have pipelines or may not. Few have waterways that transport hazardous materials. Most do not have any major air exposures but may have a local flying service that does aerial spraying of pesticides.

However, all of them have some form of highway transportation and/or fixed storage of hazardous materials within their community. Rural departments often struggle just to maintain equipment and personnel to respond to fires and medical calls and many do not have a hazardous materials response capability beyond the awareness level. I am convinced that there may even be some who do not have awareness-level training even though it is required by the Occupational Safety and Health Administration (OSHA) for all first responders and recommended by the NFPA as well. I interpret that OSHA's general point of view from their regulations is to train and equip personnel for the job they are being asked to do. Unfortunately, because regulations and training have been so structured in the past, some of those with awareness-level training have been trained beyond what is really needed in their community. In other instances, they have not received nearly enough training. For example, most awareness training includes five routes of transportation, water, pipeline, air, rail, highway, and pipeline.

Not every community has all of these exposures. Why train firefighters and other responders about water, air, or rail issues if they do not have those exposures in their communities? That time could be better spent training them for the exposure they do have. Particularly, the more common hazardous materials like LPGs, hydrocarbon fuels, anhydrous ammonia, chlorine, and pesticides which are present in the vast majority of rural communities across the United States. "Rural Hazmat Awareness" is a training concept that should provide a great deal of time on LPGs and other local hazards. Given the history of past LPG incidents, there is no reason we should continue to kill and injure firefighters responding to those incidents.

Often hazardous materials that are commonplace are overlooked because they are so common (Figure 4.99). I venture to say in some cases they may not even be considered hazardous by those who work around them frequently, the general public in the community, and even some responders. Anhydrous ammonia, for example, is very common in agricultural areas and is placarded Department of Transportation (DOT) Class 2.2, Non-Flammable Compressed Gas. This is mainly because of the

Figure 4.99 Often hazardous materials that are commonplace are overlooked because they are so common.

effects of the agricultural fertilizer industry lobby not wanting anhydrous ammonia to be considered a poison gas. Anhydrous ammonia throughout the rest of the world is placarded as a poison gas as the most severe hazard. It is also flammable under certain conditions and a corrosive gas as well. Anhydrous ammonia is not considered a flammable gas by DOT because it does not meet their definition of a flammable gas. When considering common place hazardous materials, we often think we understand their hazards and characteristics when we really don't. We do not always know how they may behave under fire and other emergency conditions. Knowledge of the containers in which hazardous materials are shipped and stored in is also important as is how those containers will react under fire and other stress conditions.

Rail cars used to transport LPGs today are insulated and may not ordinarily BLEVE as quickly as uninsulated highway and fixed storage containers might. Strategy and tactics taught and used by firefighters responding to pressurized containers on fire are often misguided. The Albert City, IA propane explosion was investigated by the National Chemical Safety and Hazard Investigation Board (CSB). Following their investigation, the Board recommended that the National Propane Gas Association (NPGA) revise its videos, manuals, and other training materials to provide appropriate instructions on responding to potential tank BLEVEs.

A similar recommendation was directed to the Fire Service Institute of Iowa State University, the fire service training organization in Iowa. We need to base our response to hazardous materials on the risk involved along with a thorough evaluation of the incident conditions present. If there is little to gain in terms of life safety by our actions, then we should not risk the safety of personnel. With the exception of Waverly, TN, many of the incidents mentioned here resulted in more injuries and deaths to firefighters than civilians. Were you to study in detail the events and circumstances surrounding these incidents you would realize that there was little to gain in most cases and too much risk to response personnel just to save property. It is ok to save property, but only if it can be done without unnecessarily risking firefighter's lives.

CSB's Herrig investigation also uncovered a potentially misleading statement in the U.S. DOT's Emergency Response Guidebook (ERG). The guidebook is carried in thousands of emergency vehicles around the country, and firefighters often consult this reference when responding to hazardous material incidents. The 1996 version of the guidebook stated that responders should "always stay away from the ends of tanks" when fighting flammable liquid tank fires. This advice could give the false impression that the sides of the tank are safer in such cases.

On the advice of the CSB, DOT revised the year 2000 guidebook, which now counsels firefighters who face propane fires to "always stay away from tanks engulfed in fire." Additionally, the ERG recommends that firefighters "fight fires involving tanks from maximum distance or use unmanned hose holders or monitor nozzles." Firefighters and especially rural firefighters should be familiar with the DOT ERG and how to use it and know its capabilities and limitations. It will likely be the only information available initially, and the isolation and evacuation distances can be particularly important for responder and public protection.

Rural fire departments may be located hours away from the nearest hazardous materials response team. Even when hazmat teams do respond to rural areas, they often only send enough personnel to perform entry into a hot zone. Local firefighters should be prepared to conduct at the very least "emergency decontamination," and can be prepared to conduct technical decontamination as well. Emergency decontamination can be safely accomplished by first responding firefighters to reduce the contamination and potential damage to public victims and any response personnel who have become contaminated. Emergency decontamination can be performed without any special equipment or protective clothing. Firefighter turnouts and self-contained breathing apparatus (SCBA) protect firefighters as long as they stay out of visible product and remain upwind and uphill from contaminated persons. Hose lines from engine companies or master streams on wide fog from elevated apparatus can provide the water necessary to conduct emergency decontamination. A number of commercial devices for emergency decontamination are also available. Departments may also train and equip specialized teams to perform technical decontamination and assist responding hazmat teams or decontaminate ambulatory victims at incident scenes even before the hazmat team arrives. Setting up a technical decontamination capability is really not that expensive. Most supplies can be obtained from local merchants who may even be willing to donate materials. Personnel protective equipment would consist of Level B protection, which is a splash suit along with SCBA that most departments should already have.

Monitoring instruments could also be helpful for known chemicals that are present in a community. For example, a four-gas meter is used to determine carbon monoxide (CO), oxygen deficiency, hydrogen sulfide (usually found in confined spaces), and lower explosive limits for flammable vapors. Confined spaces may be present in some farming operations as well as commercial grain elevators often found in rural communities. Farming associations, grain elevator operators, or community groups may be willing to provide funds for monitoring instruments. There are also federal grant programs that can be used for them.

With hazmat teams several hours away in some cases, rural departments need to prepare to deal with hazardous materials that are located in or transported through their communities. Train your personnel to use the DOT ERG whenever they suspect hazardous materials are present. On drill nights, visit hazmat locations within the community and become familiar with tanks, valves, and products. Work with your Local Emergency Planning Committee (LEPC) to find and visit locations where hazardous materials exist. Train your personnel to at the very least have the capability to conduct emergency decontamination. As a department familiarize yourself with tanks and hazardous materials in them that are stored and transported through your communities. Your efforts prior to an incident may very well prevent injuries and deaths when an incident does occur in the heartland (*Firehouse Magazine*).

Understanding Common Hazardous Materials

Looking back over historical incidents that I have researched since 1841, there seems to be a pattern between technical advances, necessity, economic gain, war, and the types of hazardous materials incidents that have occurred. From 1841 through World War II, the vast majority of major hazardous materials incidents involved explosives. When oil was discovered in Pennsylvania in the late 1860s, there were periodic interruptions from the pattern with fires and explosions involving flammable petroleum products. As technology advanced in the early 1900s with transportation and the internal combustion engine, petroleum related incidents increased. Today, these incidents account for 50% of all hazardous materials responses. Following World War II, explosives incidents involving munitions, black powder, and dynamite became infrequent. Technological development brought many new hazardous materials, and that trend has continued to the present time. Dust explosions are the only incidents that have remained constant and continue to be a major problem to date.

Major hazardous materials incidents since the 1950s have involved only a small number of very common hazardous materials. They include liquefied petroleum gases (LPG), propane and butane; ammonium nitrate; anhydrous ammonia; chlorine; combustible dust; ethanol; crude oil; and petroleum fuels. Other hazardous materials that are a risk for responders will be highlighted under Special Hazard Chemicals.

Ammonium Nitrate

Hazmatology Point: *Ammonium nitrate fertilizer is classified as an oxidizer, not as an explosive; nonetheless, it is a chemical that can detonate, and there is a century's worth of history of such explosions, some of them in manufacturing operations, but many in storage or transport. A review of incidents showed that 100% of ammonium nitrate fertilizer explosions in storage or transport had a single cause – an uncontrolled fire. Thus, ammonium nitrate fertilizer explosions in storage are preventable accidents because the technology to preclude uncontrollable* *fires also has been available for a century. In the case of transport accidents, uncontrolled fires may not be avoidable. However, technologies exist that can make ammonium nitrate less likely to explode and to show greatly reduced explosion intensity, if driven to explosion. None of these safety measures were in place for this disaster. Details of necessary fire safety measures and the effectiveness and utility of existing regulations for ammonium nitrate are examined. This is important because most ammonium nitrate storage facilities in the United States are similarly inadequate in their fire safety.*

Vitenis Babrauskas

Significant Ammonium Nitrate Incidents in U.S. Since 1947

Date	Place	Incident description	FF fatalities	Civilians
April 16, 1947	Texas City, TX	Fire/explosion	27	500
August 8, 1959	Roseburg, OR	Fire/explosion	2	12
December 17, 1960	Traskwood, AR	Fire/explosion		
November 29, 1988	Kansas City, MO	Fire/explosion	6	
December 13, 1994	Port Neal, IA	Explosion		4
July 30, 2009	Bryan, TX	Fire/no explosion		
April 17, 2013	West, TX	Fire/explosion	12	3
2014	Athens, TX	Fire/no explosion		
2019	Hastings, NE	Fire/no explosion		

Hazmatology Point: *Firefighter and civilian deaths involving ammonium nitrate fires and explosions listed above have only occurred when firefighters have chosen to fight the fire for various reasons. When firefighters evacuated citizens and stayed back themselves, the ammonium nitrate just burned with no explosion. Ammonium nitrate fertilizer, chemically classified as an oxidizer, is commonly used both in agriculture and for commercial and residential turfgrass. Approximately 7 million tons are produced annually in the United States, and under normal conditions, ammonium nitrate is a relatively safe material in terms of both storage and use. Ammonium nitrate has been in the news several times over the past two decades when it was used as an oxidizer for homemade chemical explosives by domestic and foreign terrorists. Ammonium nitrate has been involved in accidents over the years and has been responsible for the deaths of emergency responders and civilians. Most recently in West, TX, a fire and explosion occurred at a retail fertilizer dealer killing 15 people, mostly emergency responders fighting the fire. Ammonium nitrate is also used as a chemical oxidizer in* *the manufacture of the commercial explosive ammonium nitrate/fuel oil mixture (ANFO), which is a chemical mixture of ammonium nitrate and #2 fuel oil.*

Ammonium nitrate, NH_4NO_3, is a strong inorganic oxidizer that can be an explosive all by itself under certain conditions. It is primarily used as an agricultural fertilizer for its nitrogen content. Ammonium nitrate is the primary component of ANFO, a commercially available explosive. Response personnel should deal with ammonium nitrate incidents with a great deal of caution. Ammonium nitrate is a colorless or white-to-gray crystal that is soluble in water. It decomposes at 210°C (392°), releasing nitrous oxide gas and ammonia. Ammonium nitrate itself does not burn but as an oxidizer supports and enhances combustion. When in contact with other combustible materials the fire hazard is increased. A fire involving ammonium nitrate in an enclosed space can lead to an explosion. Because it is an oxidizer, a fire involving ammonium nitrate can occur in the absence of atmospheric oxygen. Ammonium nitrate may explode when exposed to strong shock or high temperatures under confinement. Contaminants may increase the explosion hazard of ammonium nitrate. Organic materials, such as chlorides, and some metals, such as chromium, copper, cobalt, and nickel, can make explosions involving ammonium nitrate more severe. National Fire Protection Association (NFPA) 704 hazards for ammonium nitrate are Health 1, Fire 0, Reactivity 3, and special information OX for oxidizer. The 4-digit identification number is 1942 with an organic coating, and 2067 as the fertilizer grade. There are a number of other ammonium nitrate mixtures that have four-digit numbers; they can be found in the Department of Transportation's (DOT) Hazardous Materials Tables and in DOT Emergency Response Guidebook (ERG).

Because we are talking about the explosive hazards of ammonium nitrate, I think it is appropriate to review what an explosion actually is. Several factors must be in place for an explosion to occur involving a chemical explosive like ammonium nitrate. An explosion involving a chemical explosive is really a fast-moving fire. In simple terms, an explosive that functions via chemical reaction creates a rapidly burning fire that is made possible by the presence of a chemical oxidizer. Atmospheric oxygen does not provide enough oxygen for a chemical explosion to take place. For a fire, you need fuel, oxygen, and heat, which when combined under the right physical conditions, create an ongoing chemical chain reaction. This process continues until the fuel is consumed, the heat is reduced, the oxygen is removed, or something interrupts the chemical chain reaction. Components that allow a fire to burn rapidly enough to produce an explosion are the presence of a fuel; instead of oxygen, you need a chemical oxidizer; plus heat or some type of initiator that creates heat. Unlike fire, you need the chemical mixture to be confined for an explosion to occur. Confinement can be accomplished with the use of a piece of pipe, plastic tube, cardboard tube, or any type of substantial container that will accomplish confinement. The material itself, if in an appropriately sized volume, can provide the confinement necessary for the explosion to occur.

Several major incidents have occurred with ammonium nitrate over the years. Texas City, TX, April 16, 1947, the SS Grandcamp was at the Port taking on a load of ammonium nitrate fertilizer to be shipped to Europe as part of the rebuilding process following World War II. Approximately 17,000,000 lb (7,700 tons) of ammonium nitrate was loaded onto the ship. Also in the harbor that fateful day was the SS High Flyer located approximately 600 ft from the Grandcamp on the same dock and loaded with 2,000,000 lb (900 tons) of ammonium nitrate. By comparison, the bomb used in the bombing of the Oklahoma City Federal Building contained 5,000 lb of ammonium nitrate.

The explosion at West, TX involved approximately 60,000 lb of ammonium nitrate (Figure 4.100). At approximately 9:12 a.m., an explosion occurred within the hold of the Grandcamp. Instantly, all 27 members of the Texas City Volunteer Fire Department at the scene were killed, some bodies were disintegrated by the heat and blast pressure of the explosion. All that remained of their fire engines were piles of twisted metal. Texas City lost all but one of their firefighters and all of their apparatus in the explosion.

On November 29, 1988, at approximately 03:40, hours the Kansas City, MO Fire Department received a call for a fire at a highway construction site.

Figure 4.100 The explosion at West, TX involved approximately 60,000 lb of ammonium nitrate.

There were several explosions following the arrival of the fire department. It was reported by the Kansas City Fire Department that the first explosion involved a split load of materials in a trailer/magazine. One compartment had approximately 3,500 lb of ANFO. The rest of the contents were approximately 17,000 lb of ANFO mixture with 5% aluminum pellets. In the second trailer/magazine there were approximately 1,000 30-lb "socks" of ANFO mixture with 5% aluminum pellets. Pumper 30 was dispatched and arrived on scene at 03:52. At 04:08, 22 min after Pumper 41 arrived and approximately 16 min after Pumper 30 arrived, the magazine exploded killing all six firefighters assigned to Pumper 41 and Pumper 30.

The terrorist bombing at the Oklahoma City Federal Building involved a homemade mixture of ammonium nitrate fertilizer and fuel oil (Figure 4.101). As a result of the terror attack, 168 people, many of them children, died and another 600 were injured. No emergency responders were killed at either incident sites as the explosions occurred before their arrival, but that may not always be the case. Over 800 buildings sustained some type of damage from ground shock and blast pressure. Of the buildings damaged, 50 were demolished. Windows were broken as far as 2 miles from the blast site, and the blast was heard 50 miles away. It registered 3.5 on the open-ended Richter Scale in Denver, CO.

Explosives in the United States are regulated by the Bureau of Alcohol, Tobacco, and Firearms (ATF) in fixed storage and the DOT when in transit. Ammonium nitrate fertilizer is not regulated by the ATF because it is not an explosive. Ammonium nitrate is regulated by the Department of Homeland Security (DHS) for security purposes. DOT classifies

Figure 4.101 The terrorist bombing at the Oklahoma City Federal Building involved a homemade mixture of ammonium nitrate fertilizer and fuel oil. (From FEMA.)

ammonium nitrate as an oxidizer. The NFPA covers storage requirements for ammonium nitrate and other chemicals in their Standard, NFPA 400.

Since the bombing of the Murrah Federal Building in Oklahoma City, the DHS monitors the storage and sale of bulk ammonium nitrate fertilizer used for agricultural purposes. This oversight has led to the reduction of ammonium nitrate used by farmers and stored for sale by retailers across the country. It has also caused the retailers who still handle the product to be very careful about information regarding its presence and can only be sold to licensed farmers.

According to the Nebraska Agribusiness association, there are 31 sites in Nebraska that still store and sell ammonium nitrate fertilizer. They would not divulge who those retailers were and advised me if I happened to locate one they would likely not want to let me take photographs or provide information. As I tried to contact retailers for this book project, I found none who were willing to admit they had ammonium nitrate, even though I talked to people who said they did. Though very frustrating to me, all of this is good because it would also make it very difficult for potential terrorists or criminals to gain access to ammonium nitrate as easily as those who used it in Oklahoma City (*Firehouse Magazine*).

CASE STUDY

On April 17, 2013, a fire and subsequent explosion occurred at the West Fertilizer Company in West, TX. Firefighters from the West Volunteer Fire Department were fighting a fire at the facility when the explosion occurred. Ammonium nitrate was located in a bin inside a seed and fertilizer building on the property. The explosion registered 2.1 on a seismograph reading from Hockley, TX, 142 miles away. Fifteen people, mostly emergency responders, were killed, over 200 were injured, and 150 buildings sustained damage. Investigators confirmed that ammonium nitrate was the source of the explosion. According to the U.S. Environmental Protection Agency (EPA) there was a report of 240 tons of ammonium nitrate on the site in 2012. According to the DHS, the company had not disclosed their ammonium nitrate stock. Federal law requires that the DHS be notified whenever anyone has more than 1 ton of ammonium nitrate on hand or 400 lb if ammonium nitrate is combined with a combustible material. The fire and explosion at West, TX was investigated by the U.S. Chemical Safety Board (CSB). Listed below are some of the observations and preliminary findings following the initial investigation. For a complete listing and to monitor the investigation of the West, TX explosion, view the CSB website at www.csb.gov.

- The explosion at West Fertilizer resulted from an intense fire in a wooden warehouse building that led to the detonation of approximately 30 tons of ammonium nitrate stored inside the wooden bins. Not only were the warehouse and bins combustible, but the building also contained significant amounts of combustible seeds, which likely contributed to the intensity of the fire.
- The building lacked a sprinkler system or other systems to automatically detect or suppress fire, especially when the building was unoccupied after hours. By the time firefighters were able to reach the site, the fire was intense and out of control. The detonation occurred just 20 minutes after the first notification to the West Volunteer Fire Department.
- Although some U.S. distributors have constructed fire-resistant concrete structures for storing ammonium nitrate, fertilizer industry officials have reported to the CSB that wooden buildings are still the norm for the distribution of ammonium nitrate fertilizer across the U.S.
- No federal, state, or local standards have been identified that restrict the sitting of ammonium nitrate storage facilities in the vicinity of homes, schools, businesses, and health care facilities. In West, Texas, there were hundreds of such buildings within a mile radius, which were exposed to serious or life-threatening hazards when the explosion occurred on April 17.
- West volunteer firefighters were not made aware of the explosion hazard from ammonium nitrate stored at West Fertilizer, and were caught in harm's way when the blast occurred. NFPA recommends that firefighters evacuate from ammonium nitrate fires of "massive and uncontrollable proportions." Federal DOT guidance contained in the Emergency Response Guidebook, which is widely used by firefighters, suggests fighting even large ammonium nitrate fertilizer fires by "flooding the area with water from a distance." However, the response guidance appears to be vague since terms such as "massive," "uncontrollable," "large," and "distance" are not clearly defined. All of these provisions should be reviewed and harmonized in light of the West disaster to ensure that firefighters are adequately protected and are not put into danger protecting property alone.

Responders in agricultural communities should be aware of the types of fertilizers stored at local facilities. Under the Emergency Planning and Community Right to Know Act (EPCRA) Ammonium nitrate is not one of the extremely hazardous substances covered by EPCRA. However, fire departments have the right to information involving chemicals at a facility for the purposes of preplanning even if the chemicals are not regulated under the act. The State Emergency Response Commission (SERC) in each state and the Local Emergency Planning Committee (LEPC) may also be of assistance.

When responding to fixed or transportation incidents in or around construction sites, mining operations, or facilities that retail agricultural fertilizers, be on the lookout for explosives or chemical oxidizers such as ammonium nitrate or commercial-grade ANFO. When responding to transportation incidents, always consider the possibility of explosives or oxidizers being present. Fire is the principal cause of accidents involving explosive materials. Look for explosive signs such as placards and labels. Evacuate the area according to the distances listed in the DOT ERG's orange section. If no other evacuation information is available, a 2,000-ft minimum distance should be observed according to the NFPA *Fire Protection Handbook*. There is one rule of thumb in responding to incidents where explosives or chemical oxidizers are involved: **DO NOT FIGHT FIRES IF THE FIRE HAS REACHED THE EXPLOSIVE STORAGE AREA OR CARGO. THE SAME APPLIES TO CHEMICAL OXIDIZERS SUCH AS AMMONIUM NITRATE** (Chemical Safety Board).

CASE STUDY

SERGEANT BLUFF, IA DECEMBER 13, 1994 AMMONIUM NITRATE EXPLOSION

The Terra Industries ammonium nitrate plant in Sergeant Bluff, south of Sioux City, exploded after an equipment malfunction on December 13, 1994, killing four and injuring 18 people (Figure 4.102). Some 3,000 residents of Iowa and Nebraska were evacuated. Ammonia gas wafted off the site for 6 days. Better safety protocols and design changes are now in place, Iowa Occupational Safety and Health Administration (OSHA) Administrator Stephen Slater said on Thursday.

According to Slater, "All kinds of technologies have had huge improvements. And we haven't had any bad experiences at the plants in the 20 years since Terra. I'm knocking on wood."

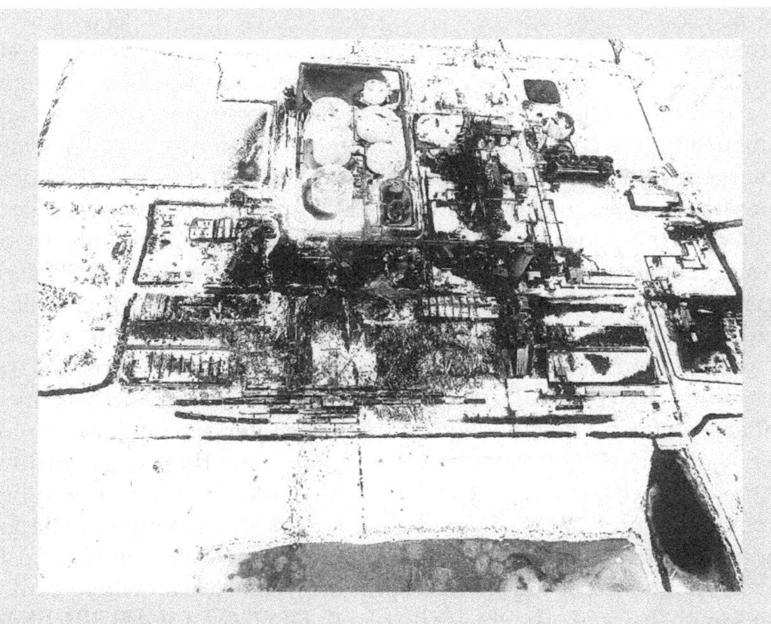

Figure 4.102 The Terra Industries ammonium nitrate plant in Sergeant Bluff, south of Sioux City, exploded after an equipment malfunction on December 13, 1994, killing 4 and injuring 18.

Slater also noted that large fertilizer plants are now subject to an extra set of safety regulations that include meticulous requirements regarding how equipment is used and replaced.

Such plants often use ammonium nitrate as a fertilizer because of its high nitrogen content, which promotes plant growth. But the chemical can also explode under certain conditions, and was the explosive used in the Oklahoma City bombing that killed 168 people in 1995. The potential for explosions has raised concerns in Iowa. Already operating in the state are the reconstructed former Terra plant, now owned by CF Industries, and a Koch Nitrogen Co. facility in Fort Dodge.

In addition, plans have been announced to expand the rebuilt Terra plant in the Port Neal complex near Sioux City and to build a second large fertilizer plant in southeastern Iowa near the Lee County town of Wever. Mitch Doherty, who lives less than a mile from the planned Lee County plant, worries that it could present problems that his community hadn't needed to be concerned about previously. "Not only do you have the facility, but you've got the trains and trucks hauling chemicals all the time," Doherty said. "You'd hope that this plant would be more modern, safer." But Doherty said Lee County supervisors and representatives of Iowa Fertilizer Co., the company planning the project, have offered only vague reassurances when asked about safety issues.

Doherty said, "That little town in Texas, it's about the same as us. You look at something like that, and think 'holy cow.'". A spokeswoman for the project's parent company, Orascom Construction Industries of Egypt, declined to comment on the Texas explosion or safety protocols it intends to have in place at the Lee County plant. CF Industries did not return calls or emails. The Deerfield, IL, company plans to invest $1.7 billion to expand the former Terra plant. There also has been talk of a possible third fertilizer plant to be built in Mitchell County in northeast Iowa near the Minnesota border. Cronus Chemical LLC has reportedly been in discussions with Iowa development officials who would provide it with $35 million in tax incentives to proceed with the project.

The U.S. EPA collects "worst-case scenario" documents from such plants and monitors their emergency preparedness plans. A spokesman for the office in Lenexa, Kan., Kris Lancaster, said because of the varying condition at different plants, his staff can't say how wide an area would be evacuated if one of the Iowa plants exploded, or how large a crater might be left. If another explosion were to occur, the Iowa Department of Natural Resources would decide how big an area to evacuate, based on air temperature, humidity, and the amount of gas released. The area would cover miles.

Barbara Lynch, who has worked in the environmental protection offices of Department of Natural Resources (DNR) for 35 years, said she knows of only one large-scale fertilizer plant explosion in Iowa, the Terra plant.

Rodney Tucker, a member of the DNR emergency response team, said all three of the large plant sites in Iowa are within coverage areas served by hazardous materials squads.

Labor Commissioner Michael Mauro said the Texas tragedy forces decision-makers at the state and federal level to refocus on safety issues (EPA).

Dynamite, Fireworks, and Other Explosives

Date	Place	Incident	FF deaths	Civilian
August 30, 1841	Syracuse, NY	Gun powder	Undocumented	Total 30
February 16, 1882	Chester, PA	Fireworks	9	8
January 15, 1895	Butte, MT	Dynamite	14	43
September 15, 1905	Avon, CT	Fuses	No fire dept.	15
April 10, 1917	Eddystone, PA	Black powder	0	50
October 4, 1918	Morgan, NJ	TNT-AN	0	100
July 10, 1926	Dover, NJ	Munitions	0	24
September 15, 1944	Hastings, NE	Munitions	0	9 military
October 16, 1948	Reno, NV	Dynamite	6	0
April 6, 1968	Richmond, IN	Black powder	0	41
June 5, 1971	Waco, GA	Dynamite	1	4
June 26, 1964	Marshalls Cr. PA	Dynamite-AN	3	2

Anhydrous Ammonia

Date	Location	Incident description	Fatalities
February 19, 1969	Crete, NE	Derailment/release	9
May 11, 1976	Houston, TX	Tanker/release	6
December 11, 1983	Houston, TX	Ice cream plant leak/ fire/explosion	0
September 17, 1984	Shreveport, LA	Cold storage leak	1 FF
January 10, 2002	Minot, ND	Derailment/release	1

Hazmatology Note: *To my knowledge, only one firefighter has died in an anhydrous ammonia fire. Even though a Class 2.2 Non-Flammable Gas according to DOT, ammonia does burn under the right conditions, usually inside a building or a confined space. Ammonia's greatest danger is toxicity. It is classified as a poison gas everywhere else in the world except for the United States. Here the agricultural fertilizer lobby keeps it under wraps so the average citizen will not get upset about the amount of poison gas being transported, stored, and used in almost every corner of our country. Emergency responders need to exercise caution when responding to anhydrous ammonia releases. SCBA and Level A chemical protective clothing is the appropriate protection from ammonia. If you do not have that protection, withdraw, evacuate citizens, and wait for a hazmat team that is equipped and trained to deal with anhydrous ammonia.*

Anhydrous ammonia bulk-storage tanks are a common sight in many communities across rural America as well as in some urban areas. Anhydrous ammonia is stored in bulk in large-capacity containers installed above or below ground. Normal above-ground storage is in uninsulated pressure storage tanks. Very large above-ground storage containers are often low pressure, refrigerated, and consequently insulated. Farmers will often transport anhydrous ammonia from fertilizer plants in trailer tanks hooked to pickup trucks or tractors. The material is not dangerous when handled properly, but if not handled carefully, it can be extremely dangerous.

Prepare for the Worst

Efforts are made by shippers, end-users, and the fertilizer industry to transport, store, and provide safe use of anhydrous ammonia. Despite the safety measures, accidents can and do occur and emergency responders need to be prepared to deal with anhydrous ammonia emergencies. Accidents may range from releases that affect only responders to those that can affect an entire community. Accidents may also involve victims who have been splashed with ammonia. Therefore, planning and training must include emergency medical people as well as other responders.

CASE STUDIES

- On January 18, 2002, a Canadian Pacific freight train derailed outside Minot, ND. Five of the cars were carrying anhydrous ammonia. Leaking ammonia killed one person and sent dozens of others to hospitals for treatment. Ten of those seeking treatment were admitted to the hospital. Some local residents were evacuated while others were asked to shelter in place. Civil Defense sirens and local radio and TV stations alerted residents (NTSB).
- In 1984, one firefighter was killed and a second was burned over 72% of his body in an anhydrous ammonia explosion and fire that occurred in Shreveport, LA. Ammonia was leaking inside a cold storage building. While firefighters were working inside in Level A chemical protection, ammonia reached an ignition source. Though it is listed as a nonflammable gas by DOT, ammonia burns inside structures and confined spaces; it is less likely to ignite out in the open. Precautions should be taken for ammonia leaks inside buildings just as for any other flammable gas (Shreveport Public Library).

- An accident occurred in the late 1990s in a cold storage building in Orrtanna, PA. Two plant maintenance workers, who were also local volunteer firefighters, were conducting routine maintenance on liquid ammonia lines within the facility. A leak occurred and the men were splashed with liquid ammonia, and they both died. Firefighting personnel responding to the incident were burned by ammonia vapors as they entered the facility in turnouts to rescue the workers (Hanover-Adams The Evening Sun, Thursday May 26, 1994).
- In Delaware County, PA, in the early 1990s, an ammonia leak occurred as workers were removing material from a cold storage building. The entire first-alarm assignment was exposed to ammonia vapors and experienced symptoms. They all had to go through decontamination and medical treatment at the scene.

Hazards

Anhydrous ammonia is classified by the U.S. DOT as a Division 2.2 Non-Flammable Gas. Unfortunately, this classification leaves two important hazards of anhydrous ammonia unidentified by the DOT placard and labeling system. Not only doe anhydrous ammonia burn under certain conditions it is also classified as a caustic (corrosive) liquid and poisonous gas in other parts of the world. U.S. manufacturers identify the hazards as flammable, toxic, and corrosive.

Anhydrous means "without water." Other chemicals also have the word anhydrous in their name and it means the same thing, without water. Anhydrous ammonia (NH_3) is a colorless liquefied gas that is free of water; for that reason, it has a high affinity for water. Thirteen hundred gallons of ammonia vapor will dissolve in just one gallon of water. It has a very sharp, intensely irritating odor – anyone in the area of a release will not want to stay!

Ammonia gas is lighter than air but also very close to the weight of air with a density of 0.6, so cold vapors or dense aerosol clouds may stay close to the ground or in low lying areas. and is easily liquefied by pressure. It has an autoignition temperature of 1,204°F and a flammable range of 16%–25%. Anhydrous Ammonia has a 4-digit identification number of 1005 and an NFPA 704 designation of Toxicity 3, Flammability 1, Reactivity 0. No Special Information.

The reason the DOT does not consider ammonia a flammable gas is the definition used for flammable gases. According to the DOT, a flammable gas has a lower explosive limit (LEL) below 13 or a flammable range of greater than 12% points. Ammonia misses the definition on both counts.

Ammonia has an LEL of 16, three points above the DOT requirement for flammable gas, and the flammable range is 10% points, not the 12 required by the DOT's definition. It does, however, burn, and has injured and killed firefighters when ignited. Normally, ammonia needs to be inside of a building or confined space to ignite. It does not usually burn outside in the open. However, during a tanker crash in Sacramento, CA, ammonia trapped under an overpass did ignite. Anhydrous Ammonia is considered to be toxic with a National Institutes of Occupational Safety and Health (NIOSH:) immediately Dangerous to Life and Health (IDLH) of 300 p.p.m. in air. According to the Centers for Disease Control (CDC) inhalation of concentrated fumes at the rate of 5,000 to 10,000 p.p.m. for short periods may be fatal. Additionally, exposure to 2,500 to 6,000 p.p.m. for 30 minutes or greater are considered dangerous to life. It has a threshold limit value (TLV) of 25 ppm in air. Inhalation of concentrated fumes may be fatal.

Necessary Protection

Responders to incidents involving anhydrous ammonia will require Level A chemical protective clothing and SCBA to protect them or to perform rescue operations. Anhydrous ammonia is also a very cold liquid, as it is released from a tank its temperature is –28°F and can cause serious thermal burns very quickly. There is no protective clothing to protect responders from the severe cold of the liquid. When released, the liquid ammonia quickly returns to the gas state at an expansion rate of 850 gallons of ammonia gas for every gallon of liquid.

Ammonia solutions that are available commercially for cleaning purposes contain anhydrous ammonia that has been dissolved in water. It still has a strong ammonia smell, but it is no longer flammable. Ammonia solutions remain corrosive and toxic, but by different routes of exposure and to a lesser degree. In higher concentrations, ammonia solutions become the liquid corrosive ammonium hydroxide, which is used in the production of lye. Anhydrous ammonia's primary uses are as an agricultural fertilizer and refrigerant in cold storage facilities. It is the oldest material used as a refrigerant, but 80% of its use is as a fertilizer. Ammonia is also an ingredient in the manufacture of gunpowder and sulfuric acid as well as illegal methamphetamines.

Theft of anhydrous ammonia for clandestine purposes has resulted in numerous leaks and injury and death to those stealing it. Valves have been left open, locks broken, and improper hoses and containers used in illegal transfer. When placed in portable propane cylinders, ammonia attacks the brass fittings and causes leaks. Stolen ammonia is transported in dangerous conditions such as in the trunks of automobiles. This can present a dangerous situation in an accident for emergency responders. In March 2000, the EPA issued a *Chemical Safety Alert for Anhydrous Ammonia Theft*.

Mild exposure to anhydrous ammonia can cause irritation to the eye, nose, and lung tissues. When ammonia is mixed with moisture in the lungs, it causes severe irritation. Ammonium hydroxide is actually produced in the lungs. Prolonged breathing can cause suffocation. The human eye is a complex organ made up of nerves, veins, and cells. The front of the human eye is covered by membranes, which resist exposure to dust and dirt. None of these can keep out anhydrous ammonia because the entire eye is about 80% water. A shot of ammonia under pressure can cause extensive, almost immediate damage to the eye. Ammonia extracts the fluid and destroys eye cells and tissues in minutes.

If you get a shot of anhydrous ammonia in your eye, the first few seconds are crucial. Immediately flush the eyes with copious amounts of water. If wearing contact lenses, remove them. Eyes doused with ammonia close involuntarily, but they must be forced open so water can flush the entire eye surface and inner lining of the eyelids. Continue to flush the eyes for at least 15 min. Get professional medical help as soon as possible to prevent permanent damage. If water is not available, fruit juice or cool coffee can be used to flush the eyes. Remove contaminated clothing and thoroughly wash the skin.

Clothing frozen to the skin by liquid ammonia can be loosened with liberal application of water. Wet clothing and body thoroughly, and then remove the clothing. Leave burns exposed to the air and do not cover with clothing or dressings. Immediately after first-aid treatment with water, take the burn victim to a physician. Do not apply salves, ointments, or oils as these can cause ammonia to burn deeper. Let a physician determine the proper medical treatment. Remove the victim to an area free from fumes if an accident occurs. If a patient is overcome by ammonia fumes and stops breathing, get the person in fresh air and give artificial respiration. The patient should be placed in a reclining position with head and shoulders elevated. Basic life support should be administered if needed.

Oxygen has been found useful in treating victims who have inhaled ammonia fumes. Administer 100% oxygen at atmospheric pressure. Any person who has been burned or overcome by ammonia should be placed under a physician's care as soon as possible. Begin irrigation with water immediately. The rescuer should use fresh water if possible.

If the incident is a farm accident, there is a requirement for water tanks for irrigation of the eyes and body on the anhydrous ammonia tank. Open water in the vicinity of an anhydrous ammonia leak may have picked up enough ammonia to be a caustic aqua ammonia solution.

This could aggravate the damage if used in the eyes or for washing burns. The victim should be kept warm, especially to minimize shock. If the nose and throat are affected, irrigate them with water continuously for at least 15 min. Take care not to cause the victim to choke. If the patient can swallow, encourage drinking lots of some type of citrus drink such as lemonade or fruit juice. The acidity will counteract some of the effects of anhydrous ammonia.

Emergency Response

Response to anhydrous ammonia emergencies can present many challenges to emergency responders. Ammonia is colorless, so there may be no visual indications of where the gas is. However, there are things to watch for. Ammonia gas will quickly turn vegetation brown. If its a time of year where the vegetation is expected to be green, then look for brown vegetation. You can also watch for animal or bird kill, which may have resulted from exposure to the released ammonia gas. Ammonia also has a strong odor; you can smell it before reaching a lethal dose. The odor threshold is 5-20 p.p.m, well below dangerous levels. However, as with all hazardous materials, responders should not be in a position to smell materials.

Firefighter turnouts do not provide protection from ammonia gas or liquid, although SCBA will protect the respiratory system. Ammonia vapors will seek out locations on the bodies of responders where there is moisture. The eyes are a major concern as they can be damaged or blindness can occur from ammonia contact. Areas in the groin and armpits are also potential moisture spots. However, firefighters in full turnouts can sweat and moisture can be present on any part of the body, depending on ambient temperatures. First responders in firefighter turnouts should avoid contact with ammonia vapors or liquid.

Because of its great affinity to water, first responders can use hose streams to decontaminate victims exposed to ammonia vapors or liquid. They can also use fog streams to dissolve ammonia gas from the air to protect victims or those in harm's way. Remember, however, that water and ammonia form ammonium hydroxide, a corrosive liquid. After victims receive emergency decontamination, efforts should be made to control the runoff.

Anhydrous ammonia can cause corrosion on metals, particularly copper, brass, or galvanized. Many parts of fire apparatus and equipment are made of brass, which can be damaged if in contact with anhydrous ammonia. Like anhydrous ammonia, LPG is stored in the same type of steel tanks. However, the fittings on the LPG tanks are brass, while those on anhydrous ammonia are black iron. Black iron fittings are ferrous metal and can spark. Brass, made of copper and zinc, is nonferrous, and will not spark.

Anhydrous ammonia stored in a tank with brass safety and control valves will eventually cause them to corrode, so the valves fail or become inoperative. Anhydrous ammonia does not have this effect on black iron. In the past, some fertilizer dealers used anhydrous farm tanks in the winter for LPG. Farmers used the tanks to run grain dryers. If the valves are not changed, the LPG can cause the brass fittings to fail, causing a fire. To compound the problem, they were not changing the placards, so a tank of LPG was transported and used with a non-flammable compressed gas placard. So be careful responding to incidents involving anhydrous farm tanks in the fall and winter. They may be carrying LPG.

In the industry, about 80% of anhydrous ammonia accidents are the result of using improper procedures, lack of training in equipment operation, or failure to follow prescribed practices. Emergency responders can avoid additional injuries and death with proper planning, training, and equipment to effectively handle anhydrous ammonia emergencies (*Firehouse Magazine*).

CASE STUDY

In March 2014, a release of anhydrous ammonia occurred at the Midwest Farmers Cooperative facility in Tecumseh, NE. An anhydrous ammonia semi-transport driven by an employee of Midwest had been positioned at the ammonia bulk plant in preparation for offloading the liquid ammonia into the bulk plant. During that offloading process an explosion reportedly occurred resulting in the release of approximately 100 lb of ammonia into the atmosphere. One employee of Midwest was killed and three other were injured, including a deputy sheriff.

This incident was investigated by the Nebraska State Fire Marshal's Office. Anhydrous ammonia is considered an extremely hazardous substance by the EPA and releases of 100 lb or more must be reported to the National Response Center (NRC) within 24 h. Common procedure at anhydrous ammonia fertilizer facilities is to vent hoses with ammonia liquid or vapor into a plastic or steel bleed-off water tank to absorb ammonia into the water. Typical plastic tanks have a 275-gallon liquid capacity with a cap on top but are not pressure tanks (Figure 4.103). By venting ammonia into water, the amount released into the air is minimal. This process while legal has its hazards and limitations. Firstly, there is a physical limit to the amount of ammonia that can be absorbed into the water. A gallon of water will only absorb 1–2 lb of ammonia before the water becomes saturated. If the water in the tank is not changed once this saturation occurs, little additional liquid or vapor can be absorbed. This may result in over-pressurization of the tank, which may result in an explosive tank failure if the tank is not vented. Secondly, when ammonia is absorbed by water a caustic chemical called ammonium hydroxide or aqua ammonia is formed. This liquid is very corrosive to the skin and can cause serious chemical burns. Initial reports

Figure 4.103 Typical plastic tanks have a 275-gallon liquid capacity with a cap on top but are not pressure tanks. (Courtesy: Kent Anderson.)

of the incident in Tecumseh indicated an explosion had occurred resulting in the release of ammonia vapors.

An investigation by the Nebraska State Fire Marshal's Office resulted in the following conclusions.

- The bleed-off valve on the liquid valve at the load-in bulkhead was accidentally left in the open position when the liquid and vapor hoses from the semi-transport were connected.
- When the primary liquid valves were opened and the transport pump was started, high-pressure liquid anhydrous ammonia was allowed to enter the plastic bleed-off water tank through the open bleed-off valve.
- This high-pressure liquid anhydrous ammonia vaporized within the plastic water container which led to an over-pressurization and eventually an explosive rupture of the container. In addition, this type of plastic container was never designed to be capable of holding pressure as a pressure vessel. It should be noted also that there is no regulatory requirement that the bleed-off water tank be a pressure vessel or that the tank be vented.
- Once the container was destroyed, the liquid anhydrous ammonia was free to vaporize and escape to the atmosphere creating a vapor cloud.
- This anhydrous ammonia release continued until an employee was able to close the bleed-off valve (Nebraska State Fire Marshal Investigation Report).

Chlorine

Date	Location	Incident	Responder	Civilian
November 15, 1975	Niagara Falls, NY	Fixed release	0	4
February 27, 1978	Youngstown, FL	Derail release	0	4
November 10, 1979	Mississauga, ON	Derail explosion	0	0
June 28, 2004	Macdona, TX	Bump release	0	3
March 24, 2005	Graniteville, SC	Derail release	0	9

Because they are in such wide use, the hazards of common chemicals sometimes are taken for granted. Complacency can set in and improper procedures may be used by those who work with the chemicals regularly, as well as by emergency responders who deal with the materials during a release resulting in injury and death. One of these chemicals is chlorine.

Chlorine is a common hazardous material found in most communities in the United States as a gas or in compound with other chemicals that can release chlorine when in contact with water or other chemicals. It is generally transported and stored as a liquefied compressed gas and can be found in 100- to 150-lb cylinders, 1-ton containers, and railroad cars.

Chlorine (elemental symbol Cl) is a nonmetallic element, a member of the halogen family of elements with an atomic number of 17 on the Periodic Table. Other halogens include fluorine, bromine, and iodine. Chlorine was discovered in 1774 by Carl Scheele, who also discovered oxygen and several other important compounds. Scheele called his discovery "dephlogisticated marine acid." Chlorine has an atomic weight of 35.453 and is a greenish-yellow diatomic gas with a pungent irritating odor, but does not exist freely as a gas in nature. Diatomic gases are elements that do not exist as a single molecule, in this case Cl, but rather as the diatomic molecule Cl_2. Other elements that are diatomic are hydrogen, nitrogen, bromine, iodine, fluorine, and oxygen. (Oxygen is often referred to as O_2 because it is a diatomic element.)

The primary source of chlorine is in the minerals halite (rock salt), sylvite, and carnallite and from chloride ion (sodium chloride) in seawater. It can be liquefied for more economical shipping, storage, and use.

Chlorine is toxic by inhalation (1 ppm in air), nonflammable, nonexplosive, and a strong oxidizer (stronger than oxygen). Because chlorine is a strong oxidizer, it will support combustion even though it is nonflammable. Chlorine has a National Institute for Occupational Safety and Health (NIOSH) immediately dangerous to life and health (IDLH) rating of 10 ppm and exposure limit time-weighted average (TWA) of 1 ppm. The OSHA ceiling for chlorine is 1 ppm. The maximum airborne concentration is 3 ppm. This is the amount to which a person could be exposed for up to 1 h without experiencing or developing irreversible or other serious health effects or symptoms that could impair the ability to take protective action.

Chlorine gas irritates the mucous membranes and the liquid burns the skin or causes irritation to the skin and may cause burning pain, inflammation, and blisters. Tissue contact with cryogenic liquid chlorine can cause frostbite injury. Chlorine's odor threshold is about 3.5 ppm, although some report that odor can be detected below the 1 ppm OSHA ceiling and TWA. Short-term exposure to low concentrations of chlorine (1–10 ppm) can result in a sore throat, coughing, and eye and skin irritation. After a few breaths at 1,000 ppm, chlorine can be fatal. Exposures to chlorine should not exceed 0.5 ppm (an eight-hour time-weighted average over a 40-hour week).

Chlorine is not known to cause cancer. Reproductive and developmental effects are not known or documented. Chlorine has a boiling point of 29°F, a freezing point of –150°F, a gas density of 2.5 (making it heavier than air), a specific gravity of 1.56 (heavier than water), and a vapor pressure of 5,168 mm Hg at 68°F. The vapor pressure of chlorine is 53.1 psi at 32°F and 112.95 psi at 77°F.

Chlorine is slightly soluble in water and reacts with a variety of other chemicals, including aluminum, arsenic, gold, mercury, selenium, tellurium, tin, and titanium. Carbon steel ignites near 483°F in contact with chlorine. It also reacts with many organic materials creating violent or explosive results. Chlorine reacts violently with acetylene, ether, turpentine, ammonia, fuel gas, hydrogen, and finely divided materials. Chlorine is placarded and labeled as a Division 2.3 Poison Gas in transportation and OSHA-mandated fixed storage. Nonbulk containers also have the corrosive label displayed. Chlorine has a 4-digit identification number of 1017 and an NFPA 704 designation of Toxicity 3, Flammability 0, Reactivity 0. and Special Information OX (oxidizer).

Chlorine was used during World War I as a choking (pulmonary) agent. On April 22, 1915, the German army released a large cloud of chlorine at Ypres, France, resulting in the deaths of 5,000 Allied soldiers and injury to 10,000 more. It could also be a potential weapon for terrorists because of its common use and availability. As a result of the military use of chlorine, much data is available about human exposure and the expected effects, both long and short term. Release of chlorine from containers as an act of terrorism could be very effective in killing hundreds, if not thousands, of people.

One of the primary uses of chlorine around the world is in the chlorination of drinking water and treatment of sewage (Figure 4.104). It is also widely used in swimming pools. Chlorine is used in the production of paper products as bleach, and in dyestuffs, textiles, petroleum products, medicines, antiseptics, insecticides, food, solvents, paints, plastics, and many other consumer products.

Exposure to chlorine can cause various signs and symptoms depending on the amount and length of exposure. There is no available antidote for chlorine exposure. Effects can be treated and most people exposed

Figure 4.104 One of the primary uses of chlorine around the world is in the chlorination of drinking water and treatment of sewage.

who survive acute exposure will likely recover with little, if any, side effects. Listed below are potential symptoms:

- Coughing
- Chest tightness
- Burning sensation in the nose, throat, and eyes
- Watery eyes (contact with liquid can cause blindness)
- Blurred vision
- Nausea and vomiting
- Burning pain, redness, and blisters on the skin if exposed to gas and skin injury; similar to frostbite if exposed to liquid (cryogenic) chlorine
- Difficulty breathing or shortness of breath
- Fluid in the lungs

Exposure to low concentrations (1–10 ppm) is likely to result in eye and nasal irritation, sore throat, and coughing. Higher concentrations (greater than 15 ppm) are likely to result in a rapid onset of respiratory distress with airway constriction and accumulation of fluid in the lungs (pulmonary edema). Additional symptoms may include rapid breathing, blue discoloration of the skin, wheezing, rales, or hemoptysis. Pulmonary injury may progress over several hours and lung collapse can also occur. It is estimated that the lowest lethal concentration for a 30-min exposure is 430 ppm.

While these symptoms can also be present with exposure to other inhalation hazards, investigation of the site and circumstances should clear up the chemical involved in most cases.

Chlorine usually does not just appear, it has a distinctive color and odor. Examinations of containers and reports of witnesses can be helpful in positive identification. There are usually no long-term health effects from sudden exposures to chlorine vapor for those who survive. Complications can occur such as pneumonia during treatment. Chronic bronchitis can also develop in people who contract pneumonia.

While it is a gas, chlorine can cause irritation and burns on contact with the skin. Therefore, firefighter turnouts are not appropriate for chlorine exposures inside the "hot zone" of a hazardous materials incident. In the past, firefighters were known to wear firefighter turnouts with petroleum jelly covering the exposed skin. Chlorine is a poison gas and requires SCBA and full Level A chemical protective clothing for anyone knowingly going into an atmosphere where chlorine is present. OSHA allows Level B protection for unknown atmospheres, which could include chlorine, as long as air monitoring is being conducted, but as soon as it is known that chlorine is present, protection should be changed to Level A.

Generally, gases do not present a serious contamination concern because it is unlikely they will stay on chemical protective clothing. When exposed to chlorine gas, responders will need to go through a minimal decontamination reduction corridor. Liquid exposure to chlorine or compounds of chlorine may require a more extensive decontamination effort. Victims will require decontamination quickly to reduce damage to the skin and eyes. Emergency decontamination would be appropriate by first responders if done from a safe distance, avoiding vapor and runoff. Exposure of victims to gas will result in minimal contamination. Removing clothing can limit the exposure to liquid chlorine and any gas that may be trapped in the victims' clothing.

When released from a container, chlorine is most concentrated at the point of the release. As with many gases and vapors, the concentration diminishes the farther away from the source you get. Evacuation and isolation distances found in the DOT ERG are based on computer modeling of chlorine releases. Isolation (hot zone) for small spills (those from a small container or a small leak from a large container) is 100 ft. From a large container (several small containers or large leak from a large container), the recommended isolation distance is 200 ft.

Evacuation distances are categorized into day and night spills. This is because the environment tends to be more stable at night, meaning a cloud will stay together longer and travel farther before dissipating. The evacuation distance for small day or night spills is one-tenth of a mile. The evacuation distance is three-tenths of a mile for a large day spill and seven-tenths of a mile for a large night spill. Several factors influence the amount of time a cloud of gas will stay together, including temperature, humidity, and wind direction and velocity. Chlorine dissipates best in warm, windy weather (*Firehouse Magazine*).

CASE STUDIES

Chlorine is a common industrial chemical in the top 10 produced chemicals annually. It is used and transported to and through almost any community in the United States and releases do occur. On November 17, 2003, in Glendale, AZ, a leak occurred as chlorine was being loaded into a railroad car. The incident forced the evacuation of the surrounding area for hours. Fourteen people were treated for symptoms such as nausea, throat irritation, and headaches. A preliminary investigation indicated that safety devices apparently failed. A full investigation has been initiated by the U.S. CSB.

Firefighters and hazmat personnel responded to a leak at a water distribution plant in North Carolina on June 13, 2003. Chlorine alarms in a pump house alerted personnel to the leak. No one was injured as hazmat personnel entered the facility and found that a manifold connected to 150-lb cylinders of chlorine had malfunctioned. Responders were able to shut off valves on the cylinders stopping the leak. Preplanning and following proper procedures led to a successful outcome (Chemical Safety Board).

In St. Louis, MO, in August 2002, a chlorine release caused injury to 63 people, including workers and nearby residents, during the offloading of a chlorine railroad car. Once again, an automatic shutdown system failed to operate. Approximately 48,000 lb of chlorine was released (*Firehouse Magazine*).

Another incident occurred on October 3, 1998, in Sun Bay South, FL. This was the sixth release from the same facility in 3 years. Chlorine was pumped from railroad cars into the facility and five of the six releases occurred during offloading operations. The most recent incident was caused by a cap that burst. A dozen people experienced difficulty breathing, and one employee experienced a burned trachea and other injuries.

A massive leak of liquefied chlorine gas occurred on May 6, 1991, in Henderson, NV in a plant that produces chlorine gas from sodium chloride. More than 200 people sought aid at local hospitals for respiratory distress caused by inhalation of the chlorine, with 30 admitted for treatment. Several first responders and the battalion chief in-charge were overcome by chlorine at the main entrance of the plant when they responded. Over 700 people were evacuated and taken to shelters with 2,000–7,000 others taken elsewhere.

The alarm was delayed when plant employees thought they could handle the release internally. Fire department response was the result of several reports from the public of strong odor in the area. The release was caused by the failure of pipes corroded by leaking acid from a heat exchanger that ate through the pipes, resulting in the release of more than 70 tons of chlorine (*Firehouse Magazine*).

Jack Rabbit Tests Dugway Proving Grounds, UT

The Jack Rabbit Project was the brainchild of Utah Valley University (UVU). Chlorine and anhydrous ammonia tests were carried out simulating a railcar release (Figure 4.105). They were chosen because of their toxicity and frequency of release.

> ***Hazmatology Point:*** *It is interesting to note that both chlorine and anhydrous ammonia are on the Hazmatology list of common materials. They have caused issues in the past for both responders and the public. These tests verify the effectiveness of sheltering in place, which was also unintentionally verified in the Crete, NE anhydrous derailment and release in 1969.*

Monitoring was conducted at various distances from the release to determine concentrations. Measurements were taken inside and outside of vehicles and structures. Vapor cloud behavior was documented at various distances. Conclusions that are beneficial to emergency responders who may encounter a railcar release of chlorine or anhydrous ammonia are as follows:

- Emergency Response Guide Book (ERG) Initial Isolation and Public Protective Action distances are consistent with the Jack Rabbit data in both an upwind and downwind environment.

Figure 4.105 The Jack Rabbit Project was the brainchild of Utah Valley University. Chlorine and anhydrous ammonia were carried out simulating a railcar release. (Jack Rabbit, SME team, Utah Valley University.)

- Sheltering in place is the most survivable option as a primary means of public protection during such an emergency if evacuation is not possible. It is better to be inside a structure of vehicle than outside until the outside chlorine concentration drops and the danger has passed. Gas concentrations will be affected by multiple factors, primarily wind and terrain.
- Vehicles continued to be operational even when exposed to ultra-high concentrations of chlorine. Escaping a chlorine plume lateral to the wind in a vehicle is the best course of action if the public or emergency responders find themselves in that position. Photo Ionization Detectors (PID) with 11.7 eV bulbs detected chlorine with reasonable accuracy and repeatability over broad chlorine concentration ranges.
- The primary strength of predictive plume models is in their use as planning guidance and/or forecasting tools rather than as emergency response tools due to the real-time uncertainty of some essential source data. First responders need to understand the application, limitations and capabilities of the plume model they use, including the widely used ALOHA® model.
- Common urban surfaces and materials were not greatly affected, even by direct liquid exposure to chlorine. Heavy hydrocarbons dissolved and metal surfaces were immediately corroded. Electronics continued to operate after exposure, however, long term operability was erratic. No residual chlorine contamination was noted.

Finally, the UVU Team found that the application and use of a risked based response process is critical to the incident considering the container, stress/breach release, wind, exposures and environmental conditions. Find additional responder information about the tests and training at the website (Utah Valley State University).

Crude Oil

Safety at a Glance

- The properties of crude oil can vary significantly by region.
- Common varieties include light sweet crude and heavy crude.
- The vast majority of the crude oil unit trains on CSX contain light sweet crude from the Bakken region.
- The only way to know which variety of crude is in the railcar is to contact CSX at 800-232-0144.

Light Sweet Crude

- May contain flammable gases including butane, pentane, and propane.
- May contain inhalation risk from hydrogen sulfide gas.
- More volatile flash point than heavy crude.

Heavy Crude (Often from Canada)

- Likely contains inhalation risk from hydrogen sulfide gas.
- Typically lower volatility than light crude.
- Crude Oil Derailments in U.S. and Canada Since 2013

Date	Location	Cars derailed	Total crude cars
July 6, 2013	Lac-Mégantic, Quebec	63	72
November 7, 2013	Aliceville, AL	12	90
December 30, 2013	Casselton, ND	21	106
February 13, 2014	Vandergift, PA	19	120
April 20, 2014	Lynchburg, VA	13	105
February 14, 2015	Gogama, Ontario	29	100
February 16, 2015	Mount Carbon, WV	27	109
March 5, 2015	Galena, IL	21	103
March 7, 2015	Gogama, Ontario	39	84
May 6, 2015	Heimdal, ND	6	107

With the advent of new technologies, production of crude oil from shale formations in North Dakota, Colorado, Pennsylvania, and Texas has significantly increased the transportation of crude oil by rail and highway over the past several years. Correspondingly, there has been an alarming number of train derailments in the United States and Canada involving crude oil. One of the reasons the magnitude of these incidents has been so great is the number of cars of crude oil that are on these trains. Trains can have over 100 cars carrying 30,000 gallons each of crude oil to refineries. Derailments have also shown that typical tank cars used for transportation of crude oil have flaws that increase the risk to emergency responders and the public.

Most crude oil in the United States is shipped through pipelines to the refineries. There is currently limited pipeline infrastructure available to ship crude oil from some of these new oil fields to the refineries, so it has to be shipped by a combination of truck and rail. This increase in the production of crude oil in North Dakota as well as at other locations in the United States and Canada comes from developments in technology involving horizontal drilling and hydraulic fracturing. The boom in oil production in North Dakota is primarily from the discovery of the Parshall Oil Field in Mountrail County in 2006. The amount of oil moving by rail in the United States has spiked since 2009 from just over 10,000 tanker cars to over 400,000 in 2013 and continues to increase.

The Bakken oil boom has propelled North Dakota into the top ranks of oil-producing states. As recently as 2007, North Dakota ranked eighth among the states in oil production. In 2008, the state overtook Wyoming and New Mexico; in 2009 it out produced Louisiana and Oklahoma. North Dakota surpassed California in oil production in December 2011, then in March 2012 it overtook Alaska to become the number two oil-producing state in the country, exceeded only by Texas.

Petroleum or crude oil is a mixture of hydrocarbons (compounds composed of carbon and hydrogen) and other chemicals. Crude oil hydrocarbons should not be confused with refined hydrocarbons with many commercial uses in our everyday life such as propane, butane, and bitumen, which all come from the same barrel of crude oil. The color and viscosity of petroleum vary markedly from one place to another. Most petroleum is dark brown or blackish in color, but it can be green, red, or yellow.

The elemental composition of crude oil is: Carbon 83%–87%, Hydrogen 10%–14%, Nitrogen 0.1%–2%, Oxygen 0.05%–1.5%, Sulfur 0.05%–6%, and metals <0.1%. Hydrocarbons found in crude oil range from having only one carbon (methane or natural gas) to having 50 (pentacontane). Four main types of hydrocarbons are found in crude oil, paraffins 15%–60%, naphthenes 30%–60%, aromatics 3%–30%, and asphaltics the balance. Paraffins, also known as alkanes, are single-bonded hydrocarbons considered to be saturated. They contain 20–40 carbons in the compounds. Naphthenes are cyclic single-bonded hydrocarbons that end in –ane such as cyclohexane. Aromatics are an organic molecule containing a benzene ring. Asphaltics is a term associated with Asphalts, a class of solid to semi-solid hydrocarbons derived from crude oil. Other chemicals found in crude oil besides hydrocarbons include sulfur, nitrogen, oxygen, and metals. The most common metals are iron, nickel, copper, and vanadium.

Crude oil is a Class 3 Flammable Liquid under the U.S. DOT Hazard Classification System. Because it is transported in bulk, it has a UN 4-digit identification number of 1267, which appears in the center of a red flammable liquid placard or in an orange rectangle near a flammable liquid placard. NFPA 704 fixed-facility marking system hazards for crude oil are Fire-3, Health-2, and Reactivity-0. Crude oil flammable characteristics can vary greatly depending on where the crude oil is found in nature. Because it is a mixture of hydrocarbons rather than a pure compound, and mixture components can vary, it is difficult to determine the exact physical properties of any one crude oil mixture.

Boiling point, flashpoint, flammable range, vapor pressure, vapor density, and specific gravity are all important physical characteristics to know when dealing with emergencies involving flammable liquids. But because crude oil is a mixture of hydrocarbons, information on physical characteristics of a given mixture may not be available. In general, the flashpoint of crude oil is below 10°C (50°F) and the boiling point is below 35°C (95°F). Vapors from crude oil spill will likely be heavier than air and will pool in low lying areas. Response resources such as the ERG, MSDS, if available, CHEMTREC, and the NRC should be used initially to obtain information about crude oil incidents.

Crude oil from the Bakken Region of North Dakota is considered by some to be more volatile than "normal" crude oil from other areas. However, when responding to an accident involving crude oil you need to understand that it is a flammable liquid. Thus, if the crude oil is not already on fire, it will burn, and you need to take precautions like you would with any flammable liquid that has been released from its container. If a flammable liquid is already on fire, none of the physical characteristics really matter. The focus needs to turn to tactics to deal with the fire conditions.

One of the "hidden" characteristics of crude oil is that it is considered a heavy hydrocarbon. Heavy hydrocarbons are usually difficult to get to burn, but once burning, they have a high heat output and are very difficult to extinguish. Water does not work well by itself for extinguishing flammable liquid fires. The specific gravity of flammable liquids, and specifically crude oil, is less than one; crude oil is immiscible with water, so it will float on the surface of water. This may allow burning crude oil to spread. Water should be used to cool tanks, protect exposures, and extinguish exposure fires if these things can be accomplished safely. Unmanned monitors should be used for water and foam applications.

According to the NFPA, a minimum of 500 gallons of water flow per minute is required as a minimum for tank cooling operations. Effective extinguishment of crude oil fires requires the use of hydrocarbon foam concentrates and appliances for application of foam. Fire extinguishment using water requires a particular volume of water to extinguish a given volume of fire. The same is true for foam firefighting operations; it requires a certain amount of foam to extinguish a given volume of flammable liquid fire. Once extinguishment is accomplished, a blanket of foam needs to be maintained on spilled liquids to prevent re-ignition. The bigger the fire, the more foam is required. Many fire departments do not immediately have access to large quantities of foam. This will limit the ability of extinguishment until enough foam concentrate can be obtained and brought to the incident site. It is a waste of time to apply foam to a flammable liquid on fire or to lay down a foam blanket if there is not enough concentrate available to accomplish the task. Initial tactical objectives may be limited to defensive actions including protection of personnel and the public through evacuations and letting the fires burn. Offensive tactics will often place emergency response personnel at greater risk.

Trains carrying crude oil can be over a mile long with hundreds of tank cars. If a derailment occurs, thousands of gallons of the product can be released from derailed tank cars (Figure 4.106). Several possible scenarios exist when a train derailment occurs involving crude oil tank cars.

Figure 4.106 Trains carrying crude oil can be over a mile long with hundreds of tank cars.

- Pool fire; where a leak occurs, liquid pools and ignites
- Flame impingement on tanks
- Tank rupture with possible fragmentation
- Fireballs
- Flash fires; where fuel vapor ignites at a point beyond the release

Most emergency response agencies will not have immediate resources or trained personnel to safely respond to or have the capability to intervene with such an event. This type of incident will require specialized equipment and trained personnel to mitigate and these resources may be hours away. Railroads should be contacted immediately through one of the following contact numbers depending on which railroad is involved.

Railroad	Emergency telephone number
BNSF Railway	(800) 832-5452
Canadian National (CN) Railway	(800) 465-9239
Canadian Pacific (CP) Railway	(800) 716-9132
CSX Transportation	(800) 232-0144
Kansas City Southern Rail Network	(877) 527-9464
Norfolk Southern Railroad	(800) 453-2530
Union Pacific	(888) 877-7267

Figure 4.107 Crude oil is typically carried on trains utilizing "Legacy" DOT-111 and in some cases the newer CPC-1232 tank cars.

Crude oil is typically carried on trains utilizing "Legacy" DOT-111 and in some cases the newer CPC-1232 tank cars (Figure 4.107). The U.S. National Transportation Safety Board (NTSB) has flagged the DOT-111 tank car as unfit to haul crude oil and ethanol because the car is prone to puncture in the event of a derailment. CPC-1232 tank cars have half head shields installed to protect from punctures. However, the cars have been involved in derailments with resulting fires, and thermally failed in at least four derailments resulting in explosions. NTSB has further determined that both the DOT-111 and CPC-1232, which are nonthermally protected, are vulnerable to thermal failures. NTSB wants these cars to be equipped with jackets, thermal protection, and appropriately sized pressure-relief devices that allows for the release of pressure under fire conditions. NTSB believes that the 10-year deadline for retrofitting or removing from service CPC 1232 tank cars is excessively long. Crude oil production in North America is expected to increase by more than 5 million barrels per day by 2024.

According to DOT's Pipeline and Hazardous Materials Safety Administration (PHMSA), "The size, scope, and resources needed to successfully manage a crude oil rail transportation incident will overwhelm the capability of most emergency response agencies." They make the following transportation and planning recommendations for emergency response personnel for dealing with crude oil incidents.

- Responses to unit train derailments of crude oil will require specialized outside resources that may not arrive at the scene for hours; therefore, it is critical that responders coordinate their activities with the involved railroad and initiate requests for specialized resources as soon as possible.

- These derailments will likely require mutual aid and a more robust on-scene Incident Management System than responders may normally use. Therefore, pre-incident planning, preparedness, and coordination of response strategies should be considered and made part of response plans, conduct drills, and exercises that include the shippers and rail carriers of this commodity.
- Evaluate the risks of personnel intervening directly in the incident. Consider the limitations of people involved and the ability to have adequate resources available on-site (e.g., sufficient firefighting foam concentrate, water supplies, appliances, equipment, trained personnel, and technical expertise), and the ability to sustain operations for extended periods of time (hours or days).
- For nonfire spill scenarios, have the concentrations of any flammable or toxic vapors present been determined using air monitoring instruments? What are the flammability and toxicity readings? Has the need for continuous air monitoring been properly evaluated and discussed with technical specialists? Can sources of ignition be removed and/or be eliminated? Are adequate foam supplies and equipment available for vapor suppression?
- Based on the results of the hazard assessment and risk evaluation process, are there adequate resources available to respond to the scene within a reasonable timeframe so that intervention efforts will be successful? (*Firehouse Magazine*).

Hazmatology Point: *If your agency is not fully prepared and capable in terms of resources, equipment, and properly trained personnel to intervene, defensive or nonintervention strategies will likely be the preferred strategic option.*

CASE STUDIES

SHEPHERDSVILLE, KY JANUARY 16, 2007 CRUDE OIL TRAIN DERAILMENT & FIRE

Executive Summary

On January 16, 2007, at about 8:43 a.m., E.S.T northbound CSX Transportation (CSX) freight train Q502-15, traveling about 47 mph through a curve, derailed 26 of its 80 cars near Shepherdsville, KY (Figure 4.108). Twelve of the derailed cars contained hazardous materials. Three of those cars breached and released significant amounts of flammable hazardous liquids, which ignited and burned. About 500 people were evacuated from the area near the accident. No one

Figure 4.108 Freight train Q502-15, traveling about 47 mph through a curve, derailed 26 of its 80 cars near Shepherdsville, KY. (From NTSB.)

was injured during the derailment; however, 50 people and 2 emergency responders were treated at local hospitals for minor injuries related to the hazardous materials release and fire. CSX estimated the total cost associated with this accident at $22.4 million. The weather was dry and cloudy, although recent rains had left the soil well saturated. The temperature was 28°F with 14 mph winds.

Probable Cause

The NTSB determines that the probable cause of the accident was the failure of the 18th railcar to properly negotiate a curve because of the inadequate side-bearing clearance of the B-end truck assembly, likely due to a broken side-bearing wedge plate attachment bolt, which caused a wheel to climb the rail, resulting in the derailment. Contributing to the derailment was (1) the undesirable contact of the truck bolster bowl rim with the car body center plate; and (2) the hollow worn wheels on the 18th car, which further diminished the steering ability of the truck assembly. After notification from the Bullitt County E911 Emergency Call Center, the ZFD responded to the scene at about 8:47 a.m. The ZFD firefighters were supported by mutual aid from neighboring communities. ZFD responders arrived on scene in about 4 min, and soon thereafter, the additional mutual aid resources arrived.

Upon arriving on scene, firefighters reported an intense fire encompassing an area of about 35,000 ft^2. As other responders arrived on scene, they observed that several of the derailed cars within the blaze were tank cars displaying hazardous materials cargo labeling. Firefighters confirmed the contents of the railcars with the CSX conductor and initiated efforts to control and suppress the fire. A voluntary evacuation order was issued for residents within 1 mile of the accident. Because of the dense, drifting black smoke from the accident, an 8-mile section of Interstate 65 was closed from about 9:11 a.m. to about 7:54 p.m. on January 16, 2007, and vehicular traffic was detoured.

Fire suppression and hazardous material cleanup activities concluded on January 20. On January 21, the evacuation order was lifted, residents returned to their homes, and repairs to the tracks were finished. The on-scene emergency response activities were discontinued on Monday, January 22, 2007 at about 5:05 p.m. (NTSB).

LAC-MÉGANTIC, QUEBEC, CA JULY 6, 2013 TRAIN DERAILMENT CRUDE OIL FIRE

On July 6, 2013, when an unattended 74-car freight train carrying Bakken Formation crude oil rolled down a 1.2% grade from Nantes and derailed downtown, resulting in the fire and explosion of multiple tank cars. Forty-two people were confirmed dead, with five more missing and presumed dead. More than 30 buildings in the town's centre, roughly half of the downtown area, were destroyed. With all the locomotives shut down, the air compressor no longer supplied air to the air brake system. As air leaked from the brake system, the main air reservoirs were slowly depleted, gradually reducing the effectiveness of the locomotive air brakes. At 00:56, the air pressure had dropped to a point at which the combination of locomotive air brakes and hand brakes could no longer hold the train, and it began to roll down the hill toward Lac-Mégantic, just over seven miles away. A witness recalled watching the train moving slowly toward Lac-Mégantic without the locomotive lights on. The track was not equipped with track circuits to alert the rail traffic controller to the presence of a runaway train.

Gathering momentum on the long downhill slope, the train entered the town of Lac-Mégantic at high speed. The TSB's final report concluded that the train was traveling at 105 kmph (65 mph), more than triple the typical speed for that location. The rail line in this area is on a curve and has a speed limit for trains of 16 kmph (10 mph) as it is located at the west end of the Mégantic rail yard. Just before the derailment, witnesses recalled observing the train passing through the crossing at an excessive speed with no locomotive lights, "infernal" noise, and sparks being emitted from the wheels.

The unmanned train derailed in downtown Lac-Mégantic at 01:14, in an area near the grade crossing where the rail line crosses Frontenac Street, the town's main street. This location is approximately 600 m (2,000 ft) northwest of the railway bridge over the Chaudière River and is also immediately north of the town's central business district. People on the terrace at Musi-Caféa bar located next to the center of the explosions saw the tank cars leave the track and fled as a blanket of oil generated a ball of fire three times the height of downtown buildings.

Between four and six explosions were reported initially as tank cars ruptured and crude oil escaped along the train's trajectory (Figure 4.109). The heat from the fires was felt as far as 2 km (1.2 miles) away. Blazing oil flowed over the ground, it entered the town's storm sewer and emerged as huge fires towering from other storm sewer drains, manholes, and even chimneys and basements of buildings in the area. The equipment that derailed included 63 of the 72 tank cars as well as the buffer car. Nine tank cars at the rear of the train remained on the track and were pulled away from the derailment site and did not explode. Almost all of the derailed tank cars were damaged, many having large breaches. About 6 million liters

Figure 4.109 Between four and six explosions were reported initially as tank cars ruptured and crude oil escaped along the train's trajectory. (Courtesy: Franklin County, Maine, USA.)

of petroleum crude oil was quickly released; the fire began almost immediately.

Around 150 firefighters were deployed to the scene, described as looking like a "war zone." Some were called in from as far away as the city of Sherbrooke, Quebec, and as many as eight trucks carrying 30 firefighters were dispatched from Franklin County, Maine, United States (Chesterville, Eustis, Farmington, New Vineyard, Phillips, Rangeley, and Strong). The fire was contained and prevented from spreading further in the early afternoon. The local hospital went to Code Orange, anticipating a high number of casualties and requesting reinforcements from other medical centers, but they received no seriously injured patients. A Canadian Red Cross volunteer said there were "no wounded. They're all dead." Approximately 1,000 people were evacuated initially after the derailment, explosions, and fires. Another 1,000 people were evacuated later during the day because of toxic fumes. Some took refuge in an emergency shelter established by the Red Cross in a local high school.

After 20 h, the center of the fire was still inaccessible to firefighters and five pools of fuel were still burning (Figure 4.110). A special fire-retardant foam was brought from an Ultramar refinery in Lévis, aiding progress by firefighters on a Saturday night. Five of the unexploded cars were doused with high-pressure water to prevent further explosions, and two were still burning and at risk of exploding 36 h later. The train's event recorder was recovered at around

Figure 4.110 After 20 h, the center of the fire was still inaccessible to firefighters and five pools of fuel were still burning. (Courtesy: Franklin County, Maine, USA.)

15:00 the next day and the fire was finally extinguished in the evening, after burning for nearly 2 days.

Forty-two bodies were found and transported to Montreal to be identified. Thirty-nine of those were identified by investigators by late August 2013 and the 40th in April 2014.

Identification of additional victims became increasingly difficult after the August 1 end of the on-site search and family members were asked to provide DNA samples of those missing, as well as dental records. The bodies of five presumed victims were never found. It is possible that some of the missing people were vaporized by the explosions. At least 30 buildings were destroyed in the center of town, including the town's library, a historic former bank, and other businesses and houses. One hundred and fifteen businesses were destroyed, displaced, or rendered inaccessible (BBC News).

Ethanol

Date	Location	Cars derailed	Total ethanol cars
October 20, 2006	New Brighton, PA	26	Unknown
June 19, 2009	Cherry Hill, IL	19	19
October 7, 2011	Tiskilwa, IL	28	12
February 7, 2011	Arcadia, OH	28	Unknown
July 11, 2012	Columbus, OH	17	3
March 8, 2017	Providence, RI	1	1
March 10, 2017	Graettinger, IA	20	14

Safety at a Glance

- Ethanol is pure grain alcohol with a denature additive. Typically denatured with the addition of 2%–5% gasoline.
- Fire burns with a low blue flame.
- Fire may require alcohol-resistant foam.

Physical and Chemical Characteristics

Ethanol in the context of this section will be denatured ethanol manufactured and used as a motor fuel and motor fuel additive. To compare with ethanol, we will first look at more common motor fuels. Gasoline and diesel fuel, hydrocarbon products refined from crude oil, have been the primary motor fuels for over 100 years. Both are flammable and have physical and chemical characteristics that make their use, storage, and transportation hazardous. They are classified by the U.S. DOT as Class 3

Flammable Liquids and are marked with red placards and labels. Gasoline and diesel fuel are usually shipped in bulk quantities so they will have 4-digit identification numbers of 1203 and 1993, respectively, placed at the center of the placards. DOT's 2020 ERG assigns Orange Guide 128 for both gasoline and diesel fuel. Guide 128 is titled Flammable Liquids (Non-Polar/Water-Immiscible). Both gasoline and diesel fuel have these characteristics. While ethanol is in the same hazard class as gasoline and diesel fuel, we will discover that the concepts of polarity and miscibility are one of the primary differences.

While gasoline and diesel fuel have some different physical characteristics under fire conditions, they are handled much the same. Firefighters should, however, be aware that those differences exist when dealing with emergencies involving both gasoline and diesel fuel. Generally, firefighters should know how to deal with gasoline and diesel fuel in a fire or other emergency because they are so common and have been around for so long. In fact, the most common hazardous materials response in most jurisdictions involves these fuels. Gasoline and diesel fuels are mixtures of hydrocarbon compounds and other additives. For example, a common blend of gasoline contains benzene, n-butane, ethyl alcohol (ethanol), n-hexane, methyl-tertiary butyl ether (MTBE), tertiary-amyl methyl ether (TAME), toluene, 1, 2, 4, trimethylbenzene, and xylene mixed isomers.

Physical and chemical characteristics will vary depending on the mixtures, which will very likely be different from one manufacturer to another. The mixture listed above has a flashpoint of –45°F, which is the same for most blends of gasoline, and an autoignition temperature >540°F. Some sources list gasoline mixtures with ignition temperatures in the 800°F range. Diesel fuel is also a mixture made up of diesel fuel and naphthalene and other additives. Diesel fuel has a flashpoint of >125°F and an autoignition temperature of 500°F. Diesel fuel is usually harder to ignite than gasoline, but when diesel fuel does ignite, it has a greater heat output than gasoline and may be harder to extinguish. Dry chemical and foam are the most common extinguishing agents depending on the amount of fuel on fire for both gasoline and diesel fuel.

Large scale manufacture and use of ethanol as a common motor fuel in passenger and other highway vehicles are fairly new. Hazards of ethanol are not as well understood by responders as gasoline and diesel fuel may be. Gasoline and diesel fuel are primarily transported by pipeline and with highway transportation vehicles. There is some rail transportation of gasoline and diesel fuel, but it is not nearly the volume as pipelines and highway transportation. Ethanol, on the other hand, is largely transported by rail (Figure 4.111), supplemented by highway and minimal pipeline transportation. It is classified by the DOT as a Class 3 Flammable Liquid just like gasoline and diesel fuel, and is marked during transportation with red placards and labels.

Figure 4.111 There is some rail transportation of gasoline and diesel fuel, but it is not nearly the volume as pipelines and highway transportation. Ethanol, on the other hand, is largely transported by rail.

DOT places various hazardous materials with similar hazards in the same hazard class because they are dealt with in a similar manner during a transportation emergency. In bulk shipments, the red Class 3 placard for alcohols and denatured alcohol will have a 4-digit identification number of 1987 at the center of the placard. The number 1987 does not specifically identify ethanol. However, alcohols as a chemical family have the same hazards when in manufacture, storage, transportation, and use. As a result, procedures for handling them during an emergency would be much the same. DOT's 2020 ERG refers to Orange Guide 127 for alcohols and denatured alcohol. Guide 127 is titled Flammable Liquids (Polar/Water-Miscible). This is directly opposite gasoline and diesel fuel and one of the major differences between these Class 3 liquids. Flammable liquids whether gasoline or ethanol are transported in atmospheric pressure containers because they have similar physical and chemical characteristics. The point I am trying to make is that gasoline, diesel fuel, and ethanol, while different in some aspects, are quite similar in others. The bottom line is they are all flammable liquids and when on fire will behave much the same. If we understand the physical and chemical characteristics of ethanol, we will find it really isn't any more or less difficult to deal with than gasoline or diesel fuel. What is important is that we need to understand how to deal with the differences.

Ethanol is a member of the alcohol hydrocarbon derivative family of chemicals. All alcohols are flammable and toxic to some degree. Hydrocarbon derivatives get their name from the fact that they are hydrocarbons to start with and have other chemical elements added to create a new chemical that has some economic value. Hydrocarbons are made up entirely of combinations of elements hydrogen and carbon covalently bonded together with single, double, or triple bonds. Single-bonded hydrocarbons are the most stable. Double- and triple-bonded hydrocarbons are more reactive and hazardous during an emergency.

Hydrocarbon derivatives are hydrocarbons with one or more of the elements oxygen, nitrogen, fluorine, chlorine, bromine, or iodine added to a hydrocarbon compound. Ethanol along with other alcohols is single-bonded with oxygen bonded to a single hydrogen (–OH) added to the hydrocarbon compound. Ethanol is an alcohol with the –OH attached to a two-carbon backbone from the hydrocarbon ethane. It is a pure chemical where gasoline and diesel fuels are mixtures of chemicals. Mixtures that form gasoline and diesel fuel include members of the hydrocarbon family.

Alcohols including ethanol are members of a hydrocarbon derivative subgroup known as polar solvents. (Polarity is a somewhat complicated phenomenon that we do not have time to discuss here. Therefore, you will just have to trust me on this one.) Water, like alcohol, is also a polar compound. Water has a molecular weight of 18. Air has an average molecular weight of 29. With a molecular weight of 18, water should be a gas at normal temperatures and pressures. But as we know it is not, it is a liquid. It is polarity that allows water to exist as a liquid. Because water and alcohols are polar compounds, they are miscible. That is to say, they mix when placed together. (There are some experiments you can do at home that will prove this is true.) Because water and alcohol mix is one of the reasons that different types of firefighting foam are required for alcohol fires versus gasoline or diesel fires. More on that later.

Gasoline and diesel fuels, on the other hand, are not polar compounds and are immiscible in water. When gasoline or diesel fuel are spilled on waterways water and the fuels form layers and the floats on top of water. When chemicals float on the surface of water they are considered to be lighter than water. Chemical reference sources will sometimes list a specific gravity for a liquid chemical. Specific gravity is the weight of the liquid versus water. Water is given a numerical weight of 1.00. Any chemical with a specific gravity greater than 1.00 will be heavier than water and sink to the bottom if it is not miscible. Chemicals with a specific gravity less than 1.00 will be lighter than water and float on the surface if it is not miscible. This is a concept similar to the relationship of chemical vapors and air known as vapor density.

Ethanol is miscible with water and if spilled in a water source would be difficult to clean up. Polarity is one of the major differences between ethanol and gasoline and diesel fuel. Several physical characteristics of flammable liquids are important in terms of the combustion of those liquids. These include flashpoint, ignition temperature, flammable range, vapor density, and heat output. Flashpoint, boiling point, and ignition temperature of flammable liquids are affected by polarity. Compounds that are polar have a tendency to have higher boiling points and flashpoints then nonpolar compounds with similar molecular weights. Within the parameters of combustion, several things need to be in place for combustion to occur. First and most importantly, a flammable liquid must be at or above its flashpoint temperature.

Flashpoint temperature is the temperature of the liquid and not the ambient temperature around the liquid or container. This is an important factor. It could be 32° ambient temperature and a chemical could have a flashpoint of 50°F. However, if the chemical was in a black container in direct sunlight it is possible the temperature of the liquid could be above its flashpoint. It is, therefore, very important that you know the temperature of the liquid when determining if a chemical could be above its flashpoint. If the temperature of the liquid is at or above its flashpoint, then combustion may occur; provided there is an ignition source and the ignition source is at or above the ignition temperature of the liquid or the liquid is heated to its autoignition temperature (both temperatures are the same value); and the fuel and air (oxygen) mixture is within the flammable range of the liquid.

If any one of the three conditions listed above is not present, combustion cannot occur. However, flashpoint temperature is the most important of the three. For combustion to occur at all, first the liquid must be at or above its flashpoint temperature. Flammable range has to do with the proper mixture of air and fuel that it takes for combustion to occur. Gasoline and diesel fuel have narrow flammable ranges of between 3 and 10. Ethanol, as with other alcohols, has a wider flammable range from 3.3 to 19. Therefore, ethanol will burn within a greater percentage of mixtures of fuel and air than gasoline and diesel fuel. The wider the flammable range, the more likely it will be that the fuel may burn inside a container.

When comparing the physical and chemical characteristics of gasoline, diesel fuel, and ethanol, it would appear that ethanol falls somewhere between gasoline and diesel fuel in terms of its combustible characteristics. Gasoline has flashpoint and ignition temperatures of −45°F and >540°F, respectively. Diesel fuel has flashpoint and ignition temperatures of >125°F and 500°F, respectively. Ethanol has flashpoint and ignition temperatures of 61.88°F and 685°F, respectively. In terms of ignition temperature, gasoline and ethanol are closer than diesel fuel, which means the liquids must be heated to near the same temperature or the ignition source temperature would be close to ignite both liquids.

Ethanol has a lower ignition temperature than most gasoline mixtures but is higher than diesel fuel. However, the flashpoint temperatures are much different between gasoline, diesel fuel, and ethanol. When exposed to water, gasoline and diesel fuel will float on top of the water. So, the real major differences between gasoline, diesel fuel, and ethanol are polarity and miscibility. This results in different firefighting tactics in terms of foam use with ethanol. Ethanol will mix with water. For this reason, the foam used to fight fires involving gasoline, diesel fuel, and ethanol need to be different to be effective. If you use regular protein foam, aqueous film forming foam (AFFF), or fluoroprotein foam, it will not work on ethanol fires because the water from the foam will mix with the alcohol and the foam blanket will break down. When fighting ethanol fires you will need to use alcohol type or polar solvent type foams. Another possible tactic for extinguishing ethanol fires is using the issue of miscibility to your advantage. If there is enough room in a container to fill the container with water, at some point the ethanol will no longer be flammable and the fire will go out because of the percentage of water applied that mixes with the alcohol.

Another different physical characteristic between ethanol and gasoline and diesel fuel is the color of the flame. Gasoline and diesel fuel burn with a yellow/orange flame. Ethanol burns with a bluish flame that may be difficult to view under certain light conditions. Ethanol fires also give off less carbon or black smoke than gasoline and diesel fuel fires. To summarize, what we have discussed about the differences and similarities between gasoline and diesel fuel and ethanol, they are all DOT Class 3 Flammable Liquids. Gasoline and diesel fuel are nonpolar mixtures that are immiscible in water. Ethanol is a pure chemical that is polar and miscible in water. In transportation, you will likely encounter larger quantities in an accident because ethanol is largely shipped by rail. Once a fire has occurred in an accident at a plant or in storage or transportation, you need to use the proper type of foam to extinguish the fire.

The Ethanol Manufacturing Facility, Process, and Associated Hazards

Date	Location	Event
September 14, 2002	Atchison, KS	Fire
October 22, 2003	Benson, MN	Fire
March 14, 2004	Chancellor, SD	Fire
January 19, 2009	Monroe, WI	Fire
September 8, 2010	Attwater, MN	Fire
April 11, 2011	Madison, IL	Fire
September 15, 2011	Pelham, GA	Fire
November 7, 2011	Nevada, IA	Fire
January 9, 2013	Peoria, IL	Explosion

Ethanol (ethyl alcohol) is the alcohol used to create beer, wine, and other alcoholic beverages. Ethanol is also produced as an alternative fuel to gasoline. Ethanol is a renewable energy source that can help reduce the U.S.

dependence on foreign oil imports. In the United States, over 9 billion gallons of ethanol fuel are produced annually. Ethanol is typically produced from corn or sugarcane. The United States is the largest producer of ethanol in the world and primarily uses corn.

Ethanol fuel does have a few disadvantages; it has a lower energy density than gasoline, so a tank of ethanol fuel will not go as far as a tank of gasoline, and ethanol fuel can be more difficult to burn in very cold temperatures. Ethanol fuel is also used as an oxygenate additive to gasoline. In the past, the chemical MTBE was used for this purpose. MTBE is hazardous and very harmful to the environment. Ethanol fuel can provide the same function without the negative effects to the environment. Oxygenating gasoline boosts the octane quality, enhances combustion, and reduces carbon monoxide emissions. This practice sees more use during winter months than it does during the summer.

As a by-product of ethanol production, the plant also produces 320,000 tons of dried distiller's grains with solubles (DDGS) each year (Figure 4.112). DDGS is not only a high-value livestock feed primarily for cattle but is also used for poultry and hogs. Three pounds of corn produces 1 lb of alcohol, 1 lb of CO_2 (released into the atmosphere), and 1 lb of distiller's grain.

Ethanol plants may vary in size but the process of production from corn is much the same. If you were to visit more than one ethanol plant, you would find them very similar in layout and the equipment used to produce ethanol. My visit was to a plant in Fairmont, NE. I was most impressed by the safety features built into the facility and the safety attitude of the facilities employees. Additionally, the plant was very clean. Cleanliness is necessary for the fermentation process to work properly.

Figure 4.112 As a by-product of ethanol production, the plant also produces 320,000 tons of DDGS each year.

Ethanol production begins with the delivery of the raw material corn to the facility. Much of it is trucked in from local producers (Figure 4.113). The day I visited the Fairmont facility there was a long line of 18 wheeler trucks lined up for several blocks to empty their load of corn into the enclosed receiving area at the plant. This same facility is used to load and ship the distiller's grain from the facility.

Raw grain is trucked into the facility and distiller's grain is shipped from the facility by rail. Transfer and storage of grain at the ethanol plant is like any other grain elevator operation; there is the potential for dust explosion from the grain dust. Grain is dumped into a collection pit and transferred into the storage bins at the facility (Figure 4.114). Dust control equipment is in place to reduce the chances of dust explosions. The elevator has an external leg equipped with explosion venting. The leg transfers the grain to and from the storage silos. While I was visiting the Fairmont facility, the dust control equipment at the intake facility shut down and there was a significant difference between the amount of dust present when the dust equipment was not working compared to when it was working. Stored grain generally contains 2–10 lb of grain dust per ton.

Ethanol is an alcohol, which is a flammable liquid according to the U.S. DOT. It is shipped under the red flammable liquid placard with the 4-digit identification number 1987 which is used for all alcohols. Other hazardous materials located at a typical ethanol plant include anhydrous ammonia (NH_3), sulfuric acid, phosphoric acid, ferric chloride, and sulphamic acid (for cleaning tanks). Welding gases are also present in the maintenance shop area of the plant. Anhydrous ammonia and sulfuric

Figure 4.113 Ethanol production begins with the delivery of the raw material corn to the facility. Much of it is trucked in from local producers.

Figure 4.114 Grain is dumped into a collection pit and transferred into the storage bins at the facility.

acid are trucked into the plant twice weekly and stored in large tanks outside the plant. They are piped inside the facility for use in the ethanol production process. Anhydrous ammonia is a toxic corrosive gas, which is also flammable, especially when inside of a facility or an enclosed area. It is classified by the DOT as a nonflammable gas because it does not meet the DOT definition of flammable gas. Anhydrous ammonia is classified as a 2.3 toxic gas under the UN Hazard Class System. NFPA 704 classifies anhydrous ammonia as Health Hazard 3, Flammability 1, and Reactivity 0. The important things to remember about anhydrous ammonia are its toxicity and flammability as it is piped into the ethanol production facility. A leak inside the plant could result in an explosion or fire if it encounters an ignition source. Contact with the skin, particularly moisture on the skin, will cause severe burns. Contact with the eyes can cause blindness. The word anhydrous means without water. As a result, anhydrous ammonia seeks water and will affect moist areas of the body. Sulfuric acid is corrosive and toxic and one of the top industrial chemicals in terms of total production quantity in the United States. DOT classifies sulfuric acid as a Class 8 Corrosive Liquid. Phosphoric acid is also a Class 8 Corrosive Liquid. Both sulfuric and phosphoric acids are listed under the NFPA 704 classification system as Health Hazard 3, Flammable Hazard 0, and Reactivity Hazard 2. They cause severe burns on contact with skin.

Ferric chloride is corrosive and can cause burns on skin contact. It is listed by NFPA 704 as Health Hazard 3, Flammability 0, and Reactivity 2. DOT classifies ferric chloride as a Class 8 corrosive solid. Sulphamic acid is corrosive and can cause burns on contact with skin. It is classified the same as ferric chloride.

Production of ethanol begins by extracting starch from the corn. Corn is moved into the processing facility on a conveyor and placed into tanks. Enzymes are used to extract the starch from the corn and is turned into sugar. Yeast is then mixed into the sugar in 8,000-gallon fermentation tanks. It is the fermentation process that ultimately produces ethanol. Fermentation creates heat, and circulating water is used to cool the process to 91° for the yeast to work properly. Sulfuric acid is used to clean equipment and adjust the pH for optimum fermentation. Anhydrous ammonia is used for pH correction and supplies nutrients for the yeast. When finished 200 proof ethanol is produced.

The 200 proof is then denatured with 2.5% natural gasoline produced from natural gas. Denaturing is done to prevent ethanol from being used as drinking alcohol. The finished product is 190 proof and is stored in 200,000-gallon closed floating roof tanks at the facility (Figure 4.115). Production capacity at the Fairmont, NE plant is 3 million gallons and there are usually 1 million gallons on hand and at any time. There are numerous railcars on-site at any given time, each with a capacity of 30,000 gallons of ethanol. Railcars are filled at a specially designed loading rack. Piping in the plant is painted to indicate what is in the pipes. Yellow pipes carry chemicals. Red pipes are 120 psi steam; silver pipes are process lines with 185°F liquids. Green pipes carry clean water. Leaking or ruptured pipes can be hazardous to emergency responders so caution should be exercised in response to pipe leaks or leaks encountered during other emergencies. Floor grates are located throughout the process area with sump pumps to route ethanol spills back into the process. The plant is

Figure 4.115 The finished product is 190 proof and is stored in 200,000-gallon closed floating roof tanks at the facility.

kept very clean, almost food process clean. Cleanliness is necessary for the fermentation process to work properly.

Dryers are located in the Energy Center building and are used to dry grain. These are powered by natural gas and produce temperatures of 800°F–900°F during normal operations. Thermal oxidizers in the same building produce temperatures of 15°F–1,600°F. Thermal oxidizers are used to burn off exhaust from the drying process to prevent odor. Steam is also a by-product of the thermal oxidizer process. Carbon dioxide and heat are the only by-products of ethanol plant operation that are released into the atmosphere. Production of ethanol produces approximately 21% less CO_2 than the production of gasoline.

Emergency responders with ethanol plants in their jurisdictions may want to become acquainted with management and safety personnel at the plant. Knowledge of the characteristics of ethanol and other hazardous materials at the facility could also be helpful. Learn the plant layout and type of fire protection equipment available. Understand the production process for ethanol and know where hazardous materials and areas are located at the facility. Proper knowledge and preplanning of the facility can make responding to an emergency safer and more efficient.

Emergency Response to Ethanol Spills and Fires

Emergency responders have been dealing with spills and fires involving gasoline and diesel fuel for over 150 years. I think it is safe to say that most responders understand the hazards of gasoline and diesel fuel as motor fuels and in some cases find dealing with them pretty routine. On the other hand, ethanol and its blends are fairly new in terms of use as a motor fuel and likely to present a challenge to emergency responders. As with any hazardous material, response personnel need to be able to recognize when they are potentially dealing with an incident involving ethanol and its blends or just plain gasoline. Containers used to transport ethanol and its blends are what we call bulk containers.

Bulk containers are required to carry placards with the 4-digit identification number that identifies the product (Figure 4.116). This number can be looked up in the DOT ERG to obtain emergency information. Following placards, one of the best sources of information is the shipping paper. MSDS sheets can also be very helpful if available. There are similarities and differences between gasoline, diesel fuel, and ethanol and its blends. While all are flammable liquids according to the DOT, they do have some important differences in terms of chemical and physical characteristics.

Firefighters need to become as familiar with these as they are with gasoline and diesel fuels. Ethanol burns with a pale blue flame and may not be visible in daylight. Ethanol and blends that are less than 15%

Figure 4.116 Bulk containers are required to carry placards with the UN 4-digit identification number that identifies the product.

gasoline will burn with little or no smoke. Ethanol blends with 10% or less ethanol will burn with thick black carbon smoke like gasoline.

During the manufacturing process and to a lesser degree during storage at the manufacturing facility, ethanol is pure 190 proof grain alcohol. As a pure alcohol, ethanol is placarded by the DOT as a flammable bulk liquid and assigned the identification number of 1170. This material is also referred to as E100. When shipped from the manufacturing facility, ethanol is denatured with 2%–5% natural gasoline also known as E-98 and E95, respectively. A blend of 95% ethanol and 5% gasoline has been assigned a 4-digit identification number of 1987 for denatured alcohol or alcohol n.o.s (not otherwise specified).

Mixtures of E95 through E99 are also assigned the 1987 4-digit identification number. Additionally, E95 may utilize the 4-digit identification number 3475. Ethanol is ultimately blended with petroleum gasoline to form a motor fuel in various concentrations depending on whether it is used as an additive/oxygenator or blended motor fuel. Ethanol and gasoline mixtures are assigned the 4-digit identification number 3475 including E11–E99. E1–E10 blends are assigned the 4-digit number 1203, which is also used for gasoline. Pure ethanol, (E100), E10, E85, E95, and gasoline are all assigned an NFPA 704 designation of Flammability 3, Health 1, and Reactivity 0.

Once ethanol is blended with gasoline, the resulting blend has physical and chemical characteristics somewhere between pure ethanol and gasoline. E10 is the most common fuel blend of ethanol and gasoline and is widely available across the country at automotive service stations.

E10 (Gasohol) is 90% gasoline and 10% ethanol. Any motor vehicle can operate using E10 without any special modifications. E85 is the next most common blend of gasoline and ethanol and is used in flex-fuel vehicles that can burn gasoline or E85. E85 is 15% gasoline and 85% ethanol.

Gasoline has a flashpoint of –40°F and pure ethanol has a flashpoint of 55°F. The flashpoint of E85 is –20°F to –4°F. The lower flashpoint gasoline lowers the higher flashpoint of pure ethanol. At lower temperatures (<32), E85 vapors are more flammable than gasoline. However, at higher temperatures, E85 vapor is less flammable than gasoline because of the higher autoignition temperature of E85. Because of lower vapor pressure and lower heat of combustion, E85 is generally less of a fire risk than gasoline. Ethanol does have a much wider flammable range than gasoline, which means it will burn in a greater number of concentrations with air than gasoline.

It is possible for materials with wide flammable ranges to burn inside containers under the right conditions. The main point I want to make here is we should not get into a line of thinking that says ethanol or gasoline or blends are any more flammable than the other. The fact is all are considered Class 3 Flammable Liquids by the DOT and have a flammability of 3 on the NFPA 704 system. If there is a spill, control ignition sources and prevent fire from occurring. If spilled fuels are already on fire, understand that different types of foam may be required to extinguish fires involving gasoline, ethanol, and blends of each.

Small fires involving ethanol and its blends can be extinguished with a Class B type fire extinguisher (dry chemical). Generally, large fires involving flammable liquids are best contained and extinguished using firefighting foam (Figure 4.117). There are two basic firefighting foams, one for hydrocarbon fires and one for alcohol or polar solvent-type fires. Fires involving ethanol/gasoline mixtures with greater than 10% alcohol (E85 for example) should be treated differently than traditional gasoline fires.

DOT recommends emergency responders refer to Orange Guide 127 of the ERG when responding to incidents involving fuel mixtures known to contain or potentially contain more than 10% alcohol. Guide 127 specifies the use of alcohol-resistant foam. Ethanol mixtures above 10% are polar/water-miscible flammable liquids and will degrade the effectiveness of nonalcohol-resistant firefighting foams. Denatured alcohol fires E-95 can only be extinguished by using AR or polar solvent foams. Alcohol mixes with water because they are both polar and the conventional AFFF and other hydrocarbon foams will break down and not be effective against the alcohol fire. Gasohol (E10) fires can be extinguished by using conventional foams because E10 contains 90% gasoline. AR-AFFF can also be used to extinguish E10 fires but

Figure 4.117 Generally, large fires involving flammable liquids are best contained and extinguished using firefighting foam.

increased application rates may be required. AR-type foams need to be applied to ethanol fires using Type II gentle application techniques. Direct application to the surface of the fuel will likely be ineffective unless fuel depth is very shallow. Most responder encounters with ethanol and its blends will occur at the manufacturing facility or in highway transportation involving a MC/DOT 307/407 or a MC/DOT 306/406 atmospheric highway tanker truck or a train of multiple railcars (Figure 4.118). Ethanol has become the largest volume hazardous material shipped in the United States. It is also shipped by barge and sea-going tanker. Very little ethanol or its blends are currently transported by pipeline. E98, E95, or denatured ethanol is the most common form transported. The primary transportation mode for ethanol is the railcar. Trains containing ethanol railcars may contain large numbers of tank cars with potentially millions of gallons of the flammable liquid in any given train. Ethanol tank cars carry approximately 30,000 gallons each.

They are liquid tank cars at atmospheric pressure. Each tank car has a pressure relief valve set to go off at 75 psi overpressure inside the tank. It is, however, possible that pressure may build up inside a tank faster than the relief valve can discharge it. This may result in the tank rupturing violently. Ethanol blends with gasoline are generally not stored at fixed sites. Denatured ethanol is stored in bulk closed floating roof tanks at storage facilities. The blending process takes place at the loading rack where ethanol and gasoline are blended as they are loaded into tanker trucks. These

Figure 4.118 Most responder encounters with ethanol and its blends will occur at the manufacturing facility or in highway transportation involving an MC/DOT 306/406 atmospheric highway tanker truck or a train of multiple railcars.

vehicles carry approximately 9,200 gallons of fuel. Barges may carry from 420,000 gallons to 4.2 million gallons of fuel. Most incidents involving ethanol and its blends with gasoline occur during transportation or transfer of product.

CASE STUDIES

Since 2000 there have been reports of at least 25 incidents involving ethanol and its blends at fixed facilities and in transportation. On June 17, 2009, 18 cars of a CN Railway train containing denatured ethanol derailed and 14 of them caught fire in Cherry Valley, IL near Rockford. Of the trains 114 cars, 74 contained denatured ethanol. A civilian sitting in their vehicle at the railroad crossing was fatally burned by the fire which engulfed her vehicle. Six others were also injured and taken to area hospitals. Approximately 600 homes in the area of the derailment were evacuated. Firefighters from 26 local departments responded to the fire that was allowed to burn itself out over several days.

Emergency response information concerning ethanol or its blends with gasoline can be obtained from the U.S. DOT ERG and MSDS. Information can also be obtained by calling CHEMTREC at 800-424-9300 during an emergency (*Firehouse Magazine*).

Combustible Dust

Date	Location	Incident description	FF fatalities
Minneapolis, MN	May 2, 1878		14
Chicago, IL	August 5, 1897	7	
Pekin, IL	January 3, 1924		42
Milwaukee, WI	April 22, 1926	3	
Westwego, LA	December 22, 1977		35
Galveston, TX	December 29, 1977		6
Corpus Christi, TX	April 7, 1981		10
Bellwood, NE	April 7, 1981		2

> ***Hazmatology Point:*** *Since 1980 according to OSHA statistics, 450 dust explosions occurred in the United States killing 130 and injuring hundreds more. Dust explosions are 100% preventable with proper housekeeping and maintenance of a facility that creates combustible dust. The CSB Dust Update 2018 data is a collection of 105 incidents that were compiled by the Incident Screening department over an 11-year period from 2006 to 2017. CSB Call to Action, Combustible Dust "Our dust investigations have identified the understanding of dust hazards and the ability to determine a safe dust level in the workplace as common challenges," said CSB Interim Executive Kristen Kulinowski. "While there is a shared understanding of the hazards of dust, our investigations have found that efforts to manage those hazards have often failed to prevent a catastrophic explosion. To uncover why that is, we are initiating this Call to Action to gather insights and feedback from those most directly involved with combustible dust hazards."*

There is a subgroup of potentially explosive materials that involve a chemical reaction type of explosion sometimes referred to as combustible dust (Figure 4.119). Combustible dust include grain dust, sawdust, flour, corn starch, and others. These are, in many cases, ordinary combustible materials or other chemicals that, because of their physical size, have an increased surface area (Figure 4.120). This increased surface area exposes more of the particles to oxygen when they are suspended in air. When these materials are suspended in air, they can become explosive if an ignition source is present.

One of the major facilities where dust explosions occur is grain elevators; explosions occur when grain dust is suspended in the air in the presence of an ignition source. The primary danger area where the explosion is likely to occur within most elevators is the "leg," or the inclined conveyor, the mechanism within the elevator that moves the grain from the entry point to the storage point (Figure 4.121). For a dust explosion to

Explosive Dusts

Coal	**Crude Rubber**	**Peanut Hulls**
Cork	**Titanium**	**Aluminum**
Flour	**Cornstarch**	**Soy Protein**
Silicon	**Walnut Shell**	**Pea Flour**
Sugar	**Zirconium**	**Magnesium**

Figure 4.119 There is a subgroup of potentially explosive materials that involves a chemical reaction type of explosion sometimes referred to as combustible dust.

Surface Area Comparison

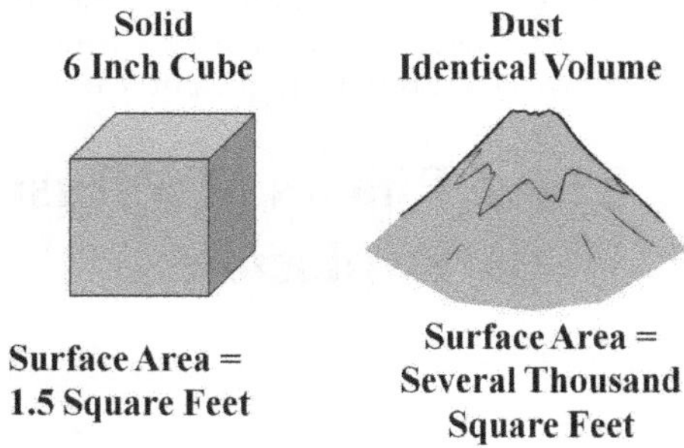

Figure 4.120 These are, in many cases, ordinary combustible materials or other chemicals that, because of their physical size, have an increased surface area.

occur, five factors must be present: an ignition source, a fuel (the dust), oxygen, a mixing of the dust and the oxygen, and confinement. The explosion will not occur unless the dust is suspended in air within an enclosure at a concentration that is above its LEL.

There are three phases in a dust explosion: initiation, primary explosion, and secondary explosion. Initiation occurs when an ignition source contacts combustible dust that has been suspended in air (Figure 4.122). This causes the primary explosion, which shakes more dust loose from the confined area and suspends it in air. The secondary explosion then occurs, which is usually the larger of the two explosions because there is more fuel present. Combustible dust may be present in many different types of facilities. Common places for combustible dust to be found are in grain elevators, flour mills, woodworking shops, and dry-bulk transport trucks.

Dusts in facilities have caused many explosions over the years that have killed and injured employees. An explosion occurred in a facility on

Figure 4.121 The primary danger area where the explosion is likely to occur within most elevators is the "leg," or the inclined conveyor, the mechanism within the elevator that moves the grain from the entry point to the storage point.

Three Phases of a Dust Explosion

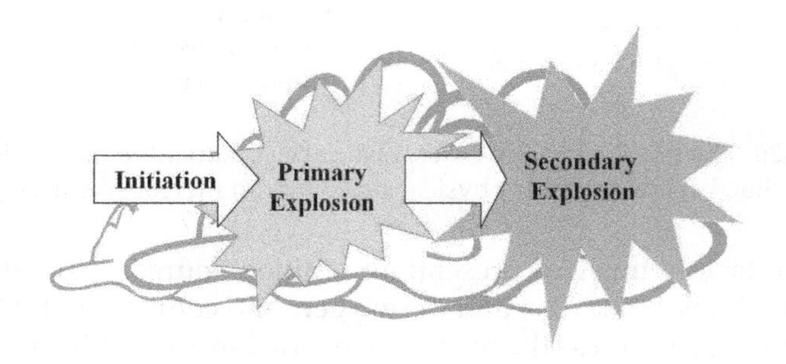

Figure 4.122 There are three phases in a dust explosion: initiation, primary explosion, and secondary explosion. Initiation occurs when an ignition source contacts combustible dust that has been suspended in air.

the East Coast that had many hazardous materials on site. At first, it was thought that one of the chemicals had exploded. The fire department and the hazmat team were called to the scene. Investigation revealed that the explosion occurred in a dust-collection system; it was a combustible dust explosion. Dust explosions can be prevented by proper housekeeping and maintenance practices at facilities where these types of dust are present (Chemical Safety Board; *Hazardous Materials Chemistry for Emergency Responders*).

Liquefied Petroleum Gases (LPG)

Date	Place	Incident	FF fatalities	Civilians
July 29, 1956	Dumas, TX	Refinery BLEVE	19	0
December 23, 1958	Brownfield, TX	Hwy Tank BLEVE	3	0
June 2, 1959	Schuylkill, PA	Hwy Tank BLEVE	1	0
June 28, 1959	Meldrim, GA	Derail BLEVE	0	23
August 13. 1963	Cleveland, OH	Fixed BLEVE	5	0
June 21, 1970	Crescent City, IL	Derail BLEVE	0	0
July 5, 1973	Kingman, AX	Fixed BLEVE	11	1
February 24, 1978	Waverly, TN	Derail BLEVE	3	13
November 10, 1979	Mississauga, ON	Derail BLEVE	0	0
December 27, 1983	Buffalo, NY	Fixed BLEVE	5	0
March 4, 1996	Weyauwega, WI	Derail Fire	0	0
October 2, 1997	Carthage, IL	Farm BLEVE	2	0
April 9, 1998	Albert City, IA	Farm BLEVE	2	0
January 30, 2007	Ghent, WV	Business BLEVE	2	2
March 12, 2007	Oneida, NY	Derail BLEVE	0	0
September 8, 2018	Tilford, SD	Home BLEVE	1	1

Propane and butane are two very common flammable liquefied compressed gases with boiling points of −42.5°C and −0.5°C, respectively. Both materials are above their boiling points under ambient temperature conditions in many parts of the country year-round. This makes the materials very dangerous when a leak or fire occurs, especially if there is flame impingement on the container. Because the materials are already above their boiling points, flame impingement, radiant heat transfer, or increases in ambient temperature can cause the materials to boil faster. Faster boiling causes an increase in vapor pressure within the container. Even though the containers are specially designed to withstand pressure and have relief valves to release excess pressure, there are limits to the pressure they can tolerate. If the pressure builds up in the container exceeds the ability of the tank to hold the pressure or the relief valve to relieve the pressure, the container will fail.

When dealing with emergencies with pressure containers and flammable gases great caution should be taken. Flame impingement on the vapor space of a container is a no-win situation. A BLEVE is going to occur, it's just a matter of time. To try to fight a fire under those conditions is to play Russian roulette. If the impingement is on the liquid space the liquid will absorb the heat for a period of time. There will also be an increase in vapor pressure within the tank because of the flame impingement. This can still be a dangerous situation if not handled properly. Copious amounts of water are needed to cool the tank down. Conditions involving the tank must be monitored constantly for changes, including liquid level, pressure coming out of the relief valve, and signs of tank failure (*Firehouse Magazine*).

CASE STUDIES

HOUSTON, TX RIMS INCIDENT

On July 2016, at 8:15 a.m. an 18 wheeler which included an MC331 Propane Tanker went around a curve and hit a concrete embankment at 55–60 mph and skidded 200 ft on the concrete roadway. Friction between the steel tank and the concrete left scrape marks on the tanker's side. Considering the mechanism of injury (damage) to the tank, hazmat team members on arrival did a visual inspection of the tank and air monitoring to determine if there was a leak. Liquid propane is transported in a noninsulated tank so the temperature of the liquid is much the same as ambient temperature. When the tank was loaded in Arkansas the ambient temperature was cool. As ambient temperature increases, so does the pressure inside the tank. Increases in heat of any kind when chemicals are involved are dangerous. Increase in heat from any source is the worst-case scenario when dealing with a hazardous material inside a container.

MC 331 tankers are equipped with temperature, pressure, and liquid level gauges. These can be helpful in determining what is going on in a tank that is not leaking. If a tank is leaking or on fire, it is too dangerous to worry about these gauges. This situation was ideal for using the gauges for important tank information. Gauges give you a visual indication there is a leak. At the time of the incident, the temperature in the container was 85°F and the tank pressure was 130. Weather reports indicated that summer temperatures in Houston would reach 110°F–115°F during the day. That temperature increase would raise the pressure inside the tank. The decision was made to keep the tank cool with a Ventura using a hose stream, which would also provide a secondary cooling effect.

This accident occurred on an elevated section of the expressway and there was no water supply. 1,000 ft of 5 in. hose was for a supply line was used to supply a ladder pipe that was used to hook the hose to, like an artificial standpipe. A tent was fashioned from the ever-innovative firefighting tool, the salvage cover. The

Figure 4.123 A tent was fashioned from the ever-innovative firefighting tool, the salvage cover. The tank was tented and hose line placed to begin the cooling process. (Courtesy: Houston Fire Department.)

tank was tented and hose line placed to begin the cooling process (Figure 4.123).

The entire well-planned process worked as intended and kept the tank cool until it could be safely offloaded and the accident scene investigated and cleared of debris. No further damage or injury occurred (*Firehouse Magazine*).

WHITE PLAINS, NY

In August 1994 in White Plains, New York, a propane tanker crashed into a house and exploded killing two people (NTSB).

KINGMAN, AZ

On July 5, 1973, a propane tank car being offloaded in Kingman, AZ, caught fire, which resulted in a BLEVE that killed 11 Kingman firefighters and 1 civilian (Figure 4.124). Another 95 persons were injured by the blast and over $1,000,000 dollars in property damage occurred to surrounding exposures. Except for one career firefighter/engineer who was severely burned but survived, those injured were mostly spectators that had gathered along historic U.S. Highway 66 to watch the incident. Most of those injured were approximately 1,000ft from the explosion and ignored police warnings to stay back. Instructors have often referenced the Kingman incident when

Figure 4.124 On July 5, 1973, a propane tank car being offloaded in Kingman, AZ caught fire, which resulted in a BLEVE that killed 11 Kingman firefighters and 1 civilian. (Courtesy: Hank Graham.)

warning emergency responders of the dangers of flame impingement on the vapor space of propane tanks. When firefighters arrived they found a 1,000-gallon propane tank involved which exploded into a 400–500 ft fireball approximately 20 min after the firefighters arrived (Figure 4.125).

Figure 4.125 When firefighters arrived they found a 1,000-gallon propane tank involved which exploded into a 400–500 ft fireball approximately 20 min after the firefighters arrived. (Courtesy: Hank Graham.)

The firefighters killed were reported to have been hit by the fireball but not burned, it is thought they died from the concussion of the explosion. Two other firefighters were injured and medivaced to the hospital and were later released (*Firehouse Magazine*).

WAVERLY, TN

The LPG explosion, which occurred in Waverly, TN on February 24th, 1978 was the high water mark of hazardous materials incidents in the United States (Figure 4.126). In terms of loss of life to emergency responders and citizens resulting from train derailments, this incident was the last in which so many people were to die. The Waverly incident also resulted in many changes in both tactics for dealing with LPG fires in containers and safety equipment on railroads.

It was Wednesday, February 24th, 1978, a typical winter day in Northwestern Tennessee. Temperatures hovered in the mid 20's with about half an inch of snow on the ground. At approximately 10:30 p.m., a Louisville and Northern (L&N) train heading from Nashville to Memphis derailed in this small community. Investigators determined that a wheel on a gondola car, overheated from a handbrake left in the applied position, broke apart East of Waverly. A wheel truck damaged by the breaking wheel managed

Figure 4.126 The LPG explosion, which occurred in Waverly, TN on February 24th, 1978, was the high water mark of hazardous materials incidents in the United States. (Courtesy: Waverly, TN Fire Department.)

to remain with the train for 7 miles before it finally came loose from the car causing the derailment. Twenty-four of the ninety-two cars of the train left the tracks in the center of downtown Waverly. Two of the derailed tank cars, which contained LPG, played a major role in the incident that unfolded over several days.

During the initial derailment, cars bumped into each other and piled on top of each other but nothing happened. No leaks, no fires, not explosions. Precautions taken including evacuations were relaxed. Railroad personnel began the process of clearing the right-of-way to get rail traffic moving again as soon as possible. They complained the mud created by the "cooling" efforts of emergency responders was making it difficult for their workers, and the hose lines were shut down but left in place. By 2:15 p.m. on the 23rd of February, the rail line had been cleared of all derailed cars. One derailed tank car #83013 (which was to BLEVE later) had been moved some 12 ft from its original resting point underneath several other rail cars (Figure 4.127). The L&N line was once again opened to limited rail traffic at 20:00 hours on the 23rd. Up to this point, no efforts had been made to deal with the LPG still in the tank cars.

At the direction of the L&N Railroad, crews were dispatched to offload the LPG tank cars and they arrived on scene about 13:00 hours on the 24th of February. By this time the sky had cleared

Figure 4.127 One derailed tank car #83013 (which was to BLEVE later) had been moved some 12 ft from its original resting point underneath several other rail cars. (Courtesy: Waverly, TN Fire Department.)

and the sun had come out raising the temperature into the mid-50's. (One of the possible contributing factors to the incident indicated by the NTSB was an increase in pressure in the tank.) Ambient temperature increases can cause the pressure in a tank to increase. Also unknown at the time was that a portion of tank #83013 had been damaged and weakened by the derailment. Keep in mind that tank cars piled up on each other or banged into each other may have sustained damage. Moving them before offloading the product can lead to further damage or catastrophic tank failure, which occurred in Waverly.

The combination of the two factors may have resulted in the BLEVE involving tank #83013. Before the offloading process was to begin, air monitoring of the area with combustible gas indicators revealed no leaks or LPG in the area. Because of the lack of catastrophic events surrounding the derailment, the evacuation was relaxed and Waverly was pretty much back to business as usual by the time offloading was to begin. In fact, persons were observed smoking in the area of the derailed LPG tanks. The fire chief, police chief, a fire crew, and two representatives of the Tennessee Civil Defense were on scene along with workers from the L&N Railroad along with personnel from the Liquid Transport Company.

Prior to initiation of the offloading process at approximately 14:58 hours someone noticed LPG vapors leaking from tank car #83013. Before anyone could react to the leaking LPG (a matter of seconds), a BLEVE occurred (Figure 4.128). The resulting fires and effects of the explosion killed five people instantly and severely burned the Waverly Fire Chief (he would die later at the hospital). The explosion compromised most of the Waverly Fire Departments on scene firefighting capability. Hoses left in place in case they were needed were shredded by the explosion, leaving no immediate means of fighting fires. Parts of the tank car, burning LPG, and other debris were scattered over a wide area.

One piece of the tank car was propelled 330 ft by the explosion. Noise and blast pressure from the explosion were felt several blocks from the scene. Numerous large buildings were set ablaze by the heat from the fireball, as well as vehicles, bystanders, and other railcars. A second LPG tank car was also set on fire by the explosion but did not BLEVE. Fortunately, flame impingement on the second LPG tank car was below the liquid level and not on the vapor space. The liquid in the tank, even though flammable, absorbed heat from flame impingement and protect the tank shell from thermal damage. In all, 16 people died from the explosion and resulting fireball

Figure 4.128 Prior to initiation of the offloading process at approximately 14:58 hours, someone noticed LPG vapors leaking from tank car #83013. Before anyone could react to the leaking LPG (a matter of seconds), a BLEVE occurred. (Courtesy: Waverly, TN Fire Department.)

and 43 people were hospitalized with injuries, and numerous others were treated as outpatients for their injuries. Sixteen buildings were totally destroyed and another twenty were damaged. The total cost of the damage in 1978 dollars was estimated at $1,800,000 (*Firehouse Magazine*).

DUMAS, TX OIL JULY 29, 1956 SHAMROCK OIL & GAS CORPORATION TANK FARM EXPLOSION

The Shamrock Oil & Gas Corporation tank farm ensuing explosions burned and shattered three other big storage tanks and blackened others nearby (Figure 4.129). Firemen attempting to cool other tanks were caught in the first explosion at 6:53 a.m. Greater loss of life was averted because many of the men had backed away from the tank because of the intense heat. The other tanks, ranging from 10,000 to 20,000 barrels in capacity, quickly caught fire from the initial blast. They contained crude oil or by-products.

Pentane and Hexane tank fires and explosions killed 19 firemen on July 29th and subsided on July 30th to a single fire in one tank. Apparently, there was no further danger. Pentane and hexane are liquids at normal temperatures and pressures. Fifteen men burned to death almost instantly when a hot wall of fire shot across the ground when the first of four tanks exploded and burned. Four others died later of horrible burns. Some 31 other persons were burned

Figure 4.129 The Shamrock Oil & Gas Corporation tank farm ensuing explosions burned and shattered three other big storage tanks and blackened others nearby. (Courtesy: Dumas, TX Fire Department.)

by the blast that shot an orange fireball thousands of feet and seared everything within a quarter-mile radius.

Big Spring Daily Herald Texas 1956-07-30

Hydrocarbon Fuels

Date	Location	Incident	FF death	Civilian
May 12, 1902	Sheridan, PA	Naphtha tank car explosion	0	21
December 25, 1928	Heidelberg, PA	Refinery fire/ explosion	2	0
July 16, 1937	Atlantic City, NJ	Gasoline tank explosion	0	0
October 20, 1944	Cleveland, OH	LNG tank explosion	0	130
May 3, 1946	Westport, CT	Black vulcanizing cement	2	2
July 22, 1947	Minot, ND	Gasoline storage tank	0	0
August 8, 1959	Kansas City, KS	Gasoline tanks explosion	5	1

(Continued)

Date	Location	Incident	FF death	Civilian
March 31, 1960	Auburn, NY	Gasoline station explosion	3	2
April 17, 1961	Philadelphia, PA	Gasoline station explosion	3	0
February 3, 1962	Houston, TX	2 gasoline tanker collision	0	2
February 7, 1968	Chicago, IL	Gasoline tanker explosion	4	5
August 17, 1975	Philadelphia, PA	Gulf refinery explosions	8	0

Author's Note: *Corrosive incidents are pretty common and usually handled well by emergency responders. I am not aware of any modern responder deaths or serious injuries caused by corrosive materials. I am including this case study of a historical incident, before hazmat, before SCBA, a corrosive incident handled by firemen.*

CASE STUDY

MILWAUKEE, WI FEBRUARY 4, 1903 ACID SPILL

Schwab Stamp & Seal Acid Spill

On February 4, 1903, an acid spill occurred at the Schwab Stamp & Seal Company. At about 2:00 p.m., an acid carboy broke open on the second floor of the facility. Employees scattered after calling the fire department. A leaking carboy of nitric acid leaked creating corrosive and toxic vapors. Cleaning up the mess was only a salvage job for the trucks and Insurance Patrol. Other companies were sent home. Chief of Department Foley and Trucks 1 & 2 located the leaking carboy and carried it outside. Then companies spent time scattering sawdust on the floor, raking it up and tossing it outside. After venting the building, Captain White of Truck 1 became ill. During the next few hours man after man complained of choking up. At 9:00 p.m. was at his bedside. Captain White died followed by Ed Hogan.

A little earlier Tom Droney of Chemical 1 had responded to another alarm. While there, he collapsed in the snow. Revived he went back to the firehouse where within a short time he also passed. By then Chief Foley was sick. Shortly after White's death, he remarked I'll bet $100, that I'll be dead by morning. He was right, he died at 4:15 a.m. The fumes had seared the men's lungs dooming them to slow suffocation (Milwaukee Fire Historical Society).

Special Hazard Chemicals

Opioids

According to the U.S. Drug Enforcement Administration (DEA), "In the last several years, U.S. Law Enforcement has witnessed a dramatic increase in the availability of dangerous synthetic opioids. A large majority of these synthetic opioids are structural derivatives of the synthetic drug "fentanyl." Fentanyl is a synthetic opioid currently listed as a Schedule II prescription drug by the DEA that mimics the effects of morphine in the human body but has a potency of 50–100 times that of morphine. Due to the high potency and availability of fentanyl, both transnational and domestic criminal organizations are increasingly utilizing these dangerous synthetic opioids as an adulterant in heroin and other controlled substances. The presence of these synthetic opioids in the illicit U.S. drug market is extremely concerning as the potency of these drugs has led to a significant increase in overdose incidents and overdose-related deaths throughout the nation.

DEA has issued the following warning to emergency responders including fire, police, (EMS) and other response personnel. "There is a significant threat to law enforcement personnel, and other first responders, who may come in contact with fentanyl and other fentanyl-related substances through routine law enforcement, emergency or life-saving activities. Because fentanyl can be ingested orally, inhaled through the nose or mouth, or absorbed through the skin or eyes, any substance suspected to contain fentanyl should be treated with extreme caution as exposure to a small amount can lead to significant health-related complications, respiratory depression, or death."

In the United States during 2016–2017, law enforcement personnel, including SWAT team members, narcotics officers, and police dogs have been exposed to fentanyl and related compounds requiring medical treatment. Fentanyl (also known as Fentanil) $C_{22}H_{28}N_2O$ IUPAC chemical name *N*-(1-(2-Phenylethyl)-4-piperidinyl)-*N*-phenylpropanamide. Carfentanil (Wildnil®) is an analogue of fentanyl with an analgesic potency 10,000 times that of morphine and is used in veterinary practice to immobilize certain large animals. It has not been approved for use in humans in any form or dose.

Illicit Uses

Fentanyl is abused for its intense euphoric effects. It can serve as a direct substitute for heroin in opioid-dependent individuals. However, fentanyl is a very dangerous substitute for heroin because it is much more potent than heroin and results in frequent overdoses that can lead to respiratory depression and death. Fentanyl patches are abused by removing the gel contents from the patches and then injecting or ingesting these contents. Patches have also been frozen, cut into pieces, and placed under the tongue or in the cheek cavity for drug absorption through the oral mucosa. Used patches are attractive to abusers as a large percentage of fentanyl remains in these patches even after a 3-day use. Fentanyl oral transmucosal lozenges and fentanyl injectables are also diverted and abused. Abuse of fentanyl initially appeared in mid-1970s and has increased in recent years. There have been reports of deaths associated with abuse of fentanyl products.

Clandestine Manufacture

"Drug incidents and overdoses related to fentanyl are occurring at an alarming rate throughout the United States and represent a significant threat to public health and safety," said DEA Administrator Michele M. Leonhart. "Often laced in heroin, fentanyl and fentanyl analogues produced in illicit clandestine labs are up to 100 times more powerful than morphine and 30–50 times more powerful than heroin." Fentanyl is extremely dangerous to law enforcement and anyone else who may come into contact with it. In the last 2 years, DEA has seen a significant resurgence in fentanyl-related seizures. According to the National Forensic Laboratory Information System (NFLIS), state and local labs reported 3,344 fentanyl submissions in 2014, up from 942 in 2013. In addition, DEA has identified 15 other fentanyl-related compounds.

Fentanyl is the most potent opioid available for use in medical treatment. Fentanyl is potentially lethal, even at very low levels. Ingestion of small doses as small as 0.25 mg can be fatal. Its euphoric effects are indistinguishable from morphine or heroin. DEA is concerned about law enforcement coming in contact with fentanyl on the streets during the course of enforcement, such as a buy-walk or buy-bust operation. The current outbreak involves not just fentanyl but also fentanyl analogues. The current outbreak is wider geographically and involves a wide array of individuals including new and experienced abusers. Carfentanil side-effects are similar to those of fentanyl, which include itching, nausea, and respiratory depression, which can be life-threatening. Carfentanil is classified as Schedule II Drug by DEA.

A 2012 study found evidence supporting the claim that during the 2002 Moscow theater hostage crisis, the Russian military made use of an aerosol form of carfentanil and another similar drug, remifentanil, to subdue Chechen hostage-takers. Its short action and easy reversibility would make it a potential agent for this purpose. Riches et al. found evidence from liquid chromatography-tandem mass spectrometry analysis of extracts of clothing from two British survivors, and urine from a third survivor, that the aerosol contained a mixture of the two anesthetics, the exact proportions of which the study was unable to determine. Previously, Wax et al.

surmised from the available evidence that the Moscow emergency services had not been informed of the use of the agent but were instructed to bring opioid antagonists. Because of the lack of information provided, the emergency workers did not bring adequate supplies of naloxone or naltrexone (opioid antagonists) to prevent complications in many of the victims, and there were subsequently over 125 confirmed deaths from both respiratory failure and aerosol inhalation during the incident. Assuming that carfentanil and remifentanil were the only active constituents, which has not been verified by the Russian military, the primary acute toxic effect to the theater victims would have been opioid-induced apnea; in this case mechanical ventilation or treatment with opioid antagonists could have been life-saving for many victims.

Naloxone/Narcan

Naloxone, sold under the brand name Narcan among others, is a medication used to block the effects of opioids, especially in overdose. Naloxone may be combined within the same pill as an opioid to decrease the risk of misuse. When given intravenously, it works within 2 min, and when injected into a muscle, it works within 5 min. The medication may also be used in the nose. The effects of naloxone last about half an hour to an hour. Multiple doses may be required as the duration of action of most opioids is greater than that of naloxone. Fentanyl has been found in busts involving heroin, morphine, oxycodone, and hydrocodone. It has also been found in cocaine and synthetic marijuana aka "Spice." Law enforcement personnel and drug-sniffing dogs are at risk from exposure to synthetic opioids. Many response organizations that may have to deal with synthetic opioids carry Narcan with them in the event of an exposure.

Weapons-Grade Narcotic

Carfentanil places law enforcement personnel at even greater risk. A chemical the size of a grain of salt can kill an officer in under 3 min. According to DEA, "It is crazy dangerous." Carfentanil is considered a weapons-grade chemical. "Terrorists could acquire it commercially as we have seen drug dealers doing," according to Andrew Weber, former U.S. Assistant Secretary of Defense for nuclear, chemical, and biological defense systems. In June 2016, 1 kg of carfentanil was seized in Vancouver. That is enough of this potent narcotic for 50 million fatal doses.

Protecting Personnel

DEA warns law enforcement officers and other emergency responders to recognize the dangers of synthetic opioids and prepare to prevent contamination. Emergency responders are advised to be vigilant with any suspected fentanyl-related narcotic.

- Exercise extreme caution.
- Don't collect samples without special training and personal protective equipment (PPE).
- Do not field test or transport samples to the office, take them directly to testing labs equipped to handle fentanyl.
- Be aware of symptoms of exposure.
- Seek immediate medical attention.
- Administer naloxone as directed.

Personnel suspected of potential exposure based on their duties should carry antidotes with them. These narcotics may look like powered cocaine, heroin, or be cut into them as a filler. Carfentanil may be found in a powder, blotter paper, tablet, patch, or spray form.

Fentanyl and analogues are extremely dangerous chemicals, but with proper awareness, training, standard operating procedures (SOPs), PPE, and antidotes, responders can deal with them safely (*Firehouse Magazine*).

Cryogenic Liquids

Cryogenic liquefied gases are very cold liquids. The U.S. Department of Transportation (DOT) defines a cryogenic liquid, sometimes referred to as a "refrigerated liquid," as any liquid with a boiling point below –130°F. Other sources list boiling points from –100°F to –200°F. If cryogenic liquids are shipped above 41 psi and have no other hazard, they are considered a compressed gas and would be placarded as a nonflammable compressed gas. Cryogenics may carry other placards such as flammable gas, poison gas, or oxidizer. If cryogenics do not have any other placardable hazard, they are not considered a hazardous material by the DOT. Materials listed under Hazard Class 2 that are shipped as liquefied gases, such as cryogenics, exhibit other hazards not indicated by the placard (Figure 4.130).

Because they have boiling points of –130°F or colder, all cryogenic liquids are above their boiling points at ambient temperatures. Liquid helium has a boiling point of –452°F below zero; it is the coldest material known. It is also the only material on earth that never exists as a solid, only as a cryogenic liquid and as a gas. Unlike propane and other liquefied gases, gases that are liquefied into cryogenics are liquefied through a process of alternating pressurization, cooling, and ultimate decompression. Therefore, they do not require pressure to keep them in the liquid state. However, if they will be in containers for long periods, they are pressurized to keep them liquefied as long as possible. Nonpressurized cryogenics are kept cold by the temperature of the liquid and the insulation around the tanks.

Figure 4.130 Materials listed under hazard Class 2 that are shipped as liquefied gases, such as cryogenics, exhibit other hazards not indicated by the placard.

The cryogenic liquefaction process begins when gases are placed into a large processing container. They are pressurized to 1,500 psi. The process of pressurizing a gas causes an increase in heat. The molecules within a container move faster, causing more collisions with each other and the walls of the container. As the molecules collide, heat is generated. For example, the top of a self-contained breathing apparatus (SCBA) bottle or an oxygen bottle becomes hot as it is being filled. This is caused by the molecules colliding in the bottle. Once the pressure of 1,500 psi is reached, the material is cooled to 32°F using ice water. Once cooled, the pressure is once again increased, this time up to 2,000 psi, again accompanied by an increase in temperature. The material is then cooled to −40°F with liquid ammonia. Once the material is cooled, all the pressure is released at once, and the resulting heat decrease turns the gases into cryogenic liquids. During the decompression process, the heat present within the container decreases as the pressure rapidly decreases.

Many of the gases found on the periodic table are extracted from the air and turned into cryogenic liquids. These include neon, argon, krypton, xenon, oxygen, and nitrogen. All the gases except oxygen are considered inert; that is, they are nontoxic, nonflammable, and nonreactive. To extract these materials from the air, the air is first turned into a cryogenic liquid. Then the liquid air is processed through a type of distillation tower where each component gas is extracted off as it reaches its own boiling point. Once extracted, the gases are then liquefied by the same process of alternate pressurization and decompression.

Other common materials that are made into cryogenics include flammable methane (LNG), hydrogen, oxidizers oxygen, fluorine, and nitric oxide.

Expansion Ratios

Cryogenic liquids have very large expansion ratios, some as much as 900 or more to one; they can form massive vapor clouds. These vapor clouds can obscure the vision of the source of the leak making the location of the source difficult to find. Vapor clouds from cryogenic liquids can travel great distances and require evacuation of the public. The visible cloud is not the total extent of the hazard. The warmer air on the outer edge of the vapor cloud causes the gas to become invisible. It is then possible to be in an oxygen-enriched atmosphere, flammable atmosphere, or in a gas that can cause asphyxiation and the gases will not be visible.

Because of the large liquid-to-gas expansion ratios, the other hazards of cryogenic liquids are magnified many times. If the cryogenic liquid is flammable or toxic, these hazards are intensified because of the potential of large gas cloud production from a small amount of liquid. As the size of the leak increases, so does the size of the vapor cloud. This means that 1 gallon of a cryogenic liquid can produce as much as 900 gallons of gas.

Cryogenic liquids can turn into gases and displace oxygen in the air, which can harm responders by simple asphyxiation – not having enough oxygen to breathe. The atmosphere contains about 21% oxygen. When the oxygen in the lungs (and ultimately in the blood) is reduced, unoxygenated blood reaches the brain and the brain shuts down. It may only be a few seconds between the first breath and collapse. Response personnel should always don SCBA before entering a confined space or other areas where asphyxiating gases may be present.

Being very cold, cryogenic liquids can cause frostbite and solidification of body parts. When the parts thaw out, the tissue is irreparably damaged. Touching uninsulated piping and valves of cryogenic liquid containers can cause the skin to stick to the metal, much like when a child's tongue is put on an ice cube tray or fence post and it becomes stuck.

There is no clothing or equipment that can be worn to protect the body from the effects of contact with a cryogenic liquid. Anything the cryogenic liquid contacts will become solidified and brittle. Gloves can prevent skin from contacting piping and valves. Body parts that have come in contact with cryogenics should be treated like frostbite. They should be flushed with water that is cool to lukewarm to limit additional tissue damage.

Cryogenic liquids are shipped and stored in special containers. On the highway, the MC 338 tanker is used to transport cryogenic liquids. The tank is usually not pressurized but is heavily insulated to keep the materials cold. A heat exchanger underneath the belly of the tank truck facilitates the offloading of product as a gas. Railroad cars are also specially designed to keep the cryogenic liquids cold inside the containers to minimize the boiling off of the gas.

Fixed storage containers of cryogenic liquids are usually tall, small-diameter tanks. These are insulated and resemble large vacuum bottles that keep the liquid cold. These containers are also under pressure to keep the material liquefied. There is a heat exchanger near the tanks. This heat exchanger is a series of metal tubes with fins around the outside. The liquids are passed through the tubes and the liquid is warmed above its boiling point and turns into a gas. The gas is then used in the facility for whatever purpose it was intended.

Valves on fixed tanks have long stems because the moisture around the valves is frozen due to the cold of the cryogenics. As this moisture is frozen, ice forms on the piping and around the valves. If the valve stem were not long, the valve would not be accessible because of the ice. Portable containers used for cryogenic liquids called Dewar flasks have an inner tank surrounded by insulation with an outer shell. The tanks have a liquid level gauge and offloading valves on the top. There is also a rupture disk in the event the container becomes over-pressurized.

Other gases, such as hydrogen, are liquefied, sometimes made into cryogenics, and placarded as flammable gases. Liquid oxygen is placarded as an oxidizer or nonflammable compressed gas. Oxygen, though nontoxic, is very reactive with hydrocarbon-based materials and acts as an oxidizer. Liquid oxygen in contact with an asphalt surface such as a parking lot or highway can create a contact explosive. Dropping an object, driving over the area, or even walking on the area can cause an explosion to occur.

Chemical Notebook:

Argon (Ar). A gaseous nonmetallic element of family eight, argon is present in the earth's atmosphere to 0.94% by volume. It is a colorless, odorless, and tasteless gas that does not combine with any other chemicals to form any compounds.

The boiling point of argon is −302°F. It is slightly soluble in water. The vapor density is 1.38, so it is heavier than air. The 4-digit identification number is 1006 as a compressed gas and 1951 as a cryogenic liquid. It is used as an inert shield in arc welding electric and specialized light bulbs (neon, fluorescent, and sodium vapor) in Geiger-counter tubes and lasers.

Carbon dioxide (CO_2). Carbon dioxide is a colorless and odorless gas. It can also be a solid (dry ice), which will undergo sublimation and turn back into carbon dioxide gas, or a cryogenic liquid. CO_2 is miscible with water and is not flammable or toxic but can be an asphyxiating gas and can displace oxygen. In 1993, two workers aboard a cargo ship were killed when a carbon dioxide fire extinguishing system discharged. The oxygen in the area was displaced by the carbon dioxide and the men were asphyxiated.

Carbon dioxide has a vapor density of 1.53, which is heavier than air. It may be shipped as a cryogenic or liquefied compressed gas. It has a 4-digit identification number of 2187 as a cryogenic and 1013 as a compressed gas.

The NFPA 704 designation is Health-3, Flammability-0, and Reactivity-0. It is used primarily in carbonated beverages and fire extinguishing systems.

When responding to fires involving liquefied gas tanks, firefighters often apply water to cool the tanks. Cryogenic liquids are already colder than water at any temperature and water will act as a superheated material, causing the cryogenic to heat and vaporize faster. This will cause pressure to build inside the tank and the tank may fail violently (*Hazardous Materials Chemistry for Emergency Responders*).

Helium (He). Helium, a gaseous nonmetallic element from family eight on the periodic table, is a colorless, odorless, and tasteless gas. It is nonflammable, nontoxic, and nonreactive. Helium has a boiling point of –452°F. It is slightly soluble in water.

Even though it is an inert gas, helium can still displace oxygen and cause asphyxiation. The vapor density is 0.1785, which is lighter than air. Helium is derived from natural gas by liquefaction of all other components. Helium has a 4-digit identification number of 1046 as a compressed gas and 1963 as a cryogenic liquid.

Krypton (Kr). Krypton is a gaseous, nonmetallic element of family eight. It is present in the earth's atmosphere to 0.000108% by volume. It is a colorless, odorless gas that is nonflammable, nontoxic, and nonreactive. It is, however, an asphyxiating gas and can displace oxygen in the air. At cryogenic temperatures, krypton exists as a white, crystalline substance with a melting point of 116° K. The boiling point of krypton is –243°F. Krypton is known to combine with fluorine at liquid nitrogen temperature by means of electric discharges or ionizing radiation to form KrF_2 or KrF_4. These materials decompose at room temperature. Krypton is slightly water-soluble. The vapor density is 2.818, which is heavier than air. The 4-digit identification number is 1056 as a compressed gas and 1970 as a cryogenic liquid. It is used in incandescent bulbs, fluorescent light tubes, lasers, and high-speed photography.

Neon (Ne). Neon is a gaseous nonmetallic element that is colorless, odorless, and tasteless. Neon is present in the earth's atmosphere at 0.0012% of normal air. It is nonflammable, nontoxic, and nonreactive and does not form chemical compounds with any other chemicals. It is, however, an asphyxiating gas and will displace oxygen in the air.

The boiling point of neon is –410°F. It is slightly soluble in water. Neon has a vapor density of 0.6964, which is lighter than air. The 4-digit identification number is 1065 when compressed and 1913 as a cryogenic liquid. Its primary uses are in luminescent electric tubes and photoelectric bulbs. It is also used in high-voltage indicators, lasers (liquid), and cryogenic research.

Nitrogen (N). Nitrogen is a gaseous, nonmetallic element. It is a colorless, odorless, and tasteless gas that makes up 78% of the air that is breathed. The boiling point of nitrogen is –320°F. It is slightly soluble in water.

Nitrogen does not burn and is nontoxic. It may, however, displace oxygen and be an asphyxiating gas. The vapor density of nitrogen is 0.96737, which makes it slightly lighter than air. The 4-digit identification number is 1066 as a compressed gas and 1977 as a cryogenic liquid. As a cryogenic liquid, the National Fire Protection Association (NFPA) 704 designation is Health-3, Flammability-0, and Reactivity-0. Nitrogen is used in the production of ammonia, cyanides, and explosives, as an inert purging agent, and as a component in fertilizers. It is usually shipped in insulated containers, insulated MC 338 tank trucks, and railroad tank cars.

Oxygen (O). Like nitrogen, oxygen is a nonmetallic elemental gas, which is the third-largest-volume industrial chemical with 53.48 billion pounds produced in 1995. Its boiling point is –297°F. It is nonflammable but supports combustion. Oxygen can explode when exposed to heat or organic materials. The vapor density of oxygen is 1.105, which makes it slightly heavier than air.

Oxygen is incompatible with: oils, grease, hydrogen, flammable liquids, solids, and gases. Its 4-digit identification number is 1072 as a compressed gas and 1073 as a cryogenic liquid. The NFPA 704 designation for liquid oxygen is Health-3, Flammability-0, and Reactivity-0. Liquid oxygen is shipped in Dewar flasks, MC 338 tank trucks, and cryogenic rail cars.

Xenon (Xe). Xenon is a gaseous, nonmetallic element from family eight. It is a colorless, odorless gas or liquid. It is a gas at standard temperatures and pressures. Xenon is nonflammable and nontoxic but is an asphyxiate and will displace oxygen in the air. The boiling point is –162°F. The vapor density is 05.987, and it is heavier than air. It is chemically unreactive but not completely inert. The 4-digit identification number is 2036 for compressed gas and 2591 for cryogenic liquid. Xenon is used in luminescent tubes, flash lamps in photography, lasers, and as an anesthetic.

Xenon compounds. Xenon combines with fluorine through a process of mixing gases, heating in a nickel vessel to 400°C, and cooling. The resulting compound is xenon tetrafluoride, XeF_4. The resulting product is composed of large colorless crystals. Compounds of xenon difluoride, XeF_2, and hexafluoride, XeF_6, can also be formed in a similar manner. The hexafluoride compound melts to a yellow liquid at 122°F and boils at 168°F.

Xenon and fluorine compounds will also combine with oxygen to form oxytetrafluoride, $XeOF_4$, which is a volatile liquid at room temperature. These compounds with fluorine must be protected from moisture to prevent the formation of xenon trioxide, XeO_3, a dangerous explosive when dried out. The solution of xenon trioxide is a stable weak acid that is a strong oxidizing agent.

Clandestine Methamphetamine Drug Labs

Clandestine drug laboratories continue to present a significant law enforcement and emergency response problem across the United States (Figure 4.131). These illegal labs contain the chemicals and equipment required to manufacture controlled substances such as methamphetamines (speed, crack, ice, glass, and crystal), phenyl-2-propanone (P2P), LSD, PCP (angel dust), MDA/MDPP (Ecstasy), methaqualone, methcathinone (cat), fentanyl, and others.

Methamphetamines are by far the most common illegal drugs manufactured in clandestine labs and will be the primary focus here. Illegal drug labs have been discovered in homes, apartments, hotel, and motel rooms, barns, restaurants, fields, vacant and abandoned buildings, storage facilities, and even mobile labs. This is a problem that concerns rural America as well as urban areas. In fact, the more remote the area, drug makers think, the less likely they will be detected by law enforcement. However, the very nature and dangers of clandestine drug labs may cause emergency responders to encounter them by accident or when something goes wrong with the chemicals involved. Other public officials or civilians may also discover clandestine drug labs.

Response personnel should become familiar with the detection of clues and hazards of the hazardous materials involved in drug lab operations. The chemicals used are themselves dangerous, and they can produce hazardous byproducts and cause fires and explosions. Use of the chemicals can often result in contamination of the area used for the drug lab, which becomes a secondary contamination hazard for response personnel. Illegal drugs can be made with "preferred" or "alternate" chemicals.

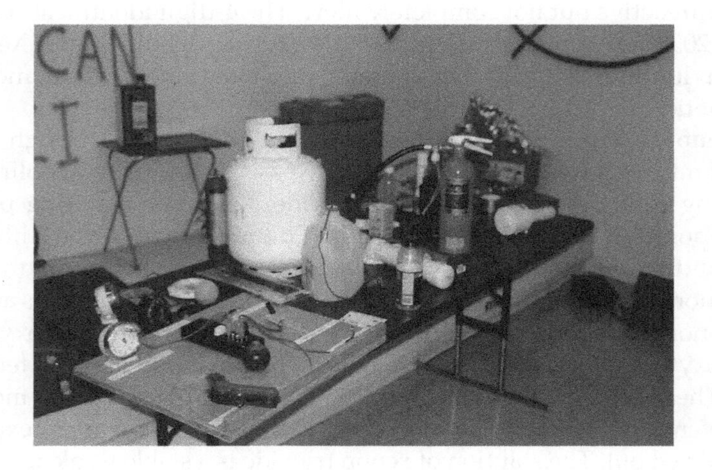

Figure 4.131 Clandestine drug laboratories continue to present a significant law enforcement and emergency response problem across the United States.

Some of these chemicals can make the operations more dangerous than others to the operators and the emergency responders.

Making illegal drugs does not require the sophistication, knowledge, or equipment necessary to manufacture chemical and biological terrorism agents. Not all chemicals associated with illegal drug manufacture are regulated and many are available from local merchants such as pharmacies, hardware stores, supermarkets, discount and convenience stores, and agricultural cooperatives. Transactions that occur in these locations involving large purchases of suspected chemicals should be reported to law enforcement. Even better, retailers should be taught to limit the amounts of materials sold, making the drug lab operators work harder to obtain the raw materials needed for drug manufacture. Though not consistent in all locations, some chemicals which may be used in illegal drug manufacture are regulated by government agencies. Manufacturers, distributors, and retailers should be aware of precursor chemicals used for illegal drug production.

Anhydrous ammonia is a common hazardous material used to make methamphetamines. It is often stolen from storage tanks on farms and commercial facilities, which may result in leaks and releases that require the response of emergency personnel. Frequently, the amounts of ammonia stolen are so small that they are not missed. Thieves then hide stolen ammonia in places where they are not expected to be found, such as the trunks of cars, inside vans, and in homes and apartments. Sometimes, thieves use portable propane tanks like those used for barbecue grills, which creates a hazard from ammonia attacking the fittings and valves that in turn may result in a release of the contents. The major problem with the theft of ammonia is that leaks often occur because valves are damaged or left open.

Detection

Response personnel must know how to recognize the hazards involved in emergencies that may involve clandestine drug labs. When a drug lab is suspected or discovered, it becomes a crime scene and law enforcement should be contacted immediately. Much like terrorist incident scenes, responders should take care to make sure the scene is safe (free of devices set up to injure response personnel) and take only those actions necessary to save lives and protect exposures.

In addition to all of the other things they should be looking for at an emergency scene, responders should be aware of the potential clues that point to an illegal drug operation. Clues to watch for include blackened out windows, burn pits, stained soil, dead vegetation, and multiple over-the-counter-drug containers. Tree kills have been found around locations of methamphetamine labs. Some 150-year-old Ponderosa Pines were killed close to a drug lab in Arizona.

Residents never putting trash out, laboratory glassware being carried into a residence, and little or no traffic during the day, but lots of traffic late at night, can also be clues to illegal drug operations. Empty containers from antifreeze, white gas, starting fluids, Freon, lye or drain openers, paint thinner, acetone or alcohol may be clues, especially if they seem out of place for the occupancy. Additional clues are anhydrous ammonia or propane tanks, ceramic or glass containers or other kitchenware with hoses or duct tape, and thermos bottles or other cold storage containers. Also suspect are respiratory masks and filters, dust masks, rubber gloves, funnels, hosing, and clamps.

Unusual odors may also be present. When residential occupancies contain odors of ammonia, solvents, chemicals, sweet, or bitter smells, they should be investigated to determine the cause. Manufacturers of illegal drugs often set up booby traps for law enforcement and other responders.

Drug Lab Chemicals

Following is a list of some of the more dangerous and in some cases common hazardous chemicals that may be found in conjunction with clandestine drug lab operations. While the presence of any one of the materials does not automatically indicate a drug operation, the type of location and numbers of chemicals present should be taken into account. The list is not meant to be comprehensive or contain response operational information. It is provided as awareness information for emergency responders. Once chemicals are located at an incident scene, they should be researched using the same methods as other hazardous materials found at a hazmat incident.

Chemical Notebook

- **Acetic acid (glacial)** is a corrosive organic acid, colorless liquid, and a solid below 62°F (glacial means solid at normal temperatures). It has a vinegar-like odor. It is used in the manufacture of phenyl-2-propanone P2P, methamphetamine, and amphetamine. Its primary hazards are corrosivity and a strong irritant. At certain concentrations it can also be flammable.
- **Acetic anhydride** is a colorless liquid with a strong vinegar-like odor. It can also be used in P2P synthesis. It is corrosive and can cause skin burns.
- **Acetone (dimethyl ketone)**, a member of the ketone family of hydrocarbon derivatives, is a volatile, highly flammable, and colorless liquid solvent with a sweet type odor. It is a common ingredient in nail polish remover and may be found in beauty salons where nail technicians work. It is not highly toxic but is a narcotic. Acetone is used in the manufacture of methamphetamines.
- **Anhydrous ammonia** is a colorless, lighter-than-air gas with a strong, pungent odor. It is toxic and can be flammable under certain

conditions, particularly inside a structure. Ammonia is readily available in rural areas, where it is used as a fertilizer and has been the target of theft for use in clandestine drug operations. Ammonia in liquid form is reacted with sodium metal that is water-reactive.

- **Benzene**, an aromatic hydrocarbon, is a colorless-to-yellow liquid with an aromatic odor (characteristic of all aromatic hydrocarbons) that is flammable, toxic, and known to cause cancer. It is a solvent used in methamphetamine production.
- **Ephedrine** is composed of odorless white crystals. It can be found in medicines available over the counter. It is one of the primary precursors used in methamphetamine production. Ephedrine is an irritant and mildly toxic.
- **Ethanol**, an alcohol hydrocarbon derivative, is a clear, colorless liquid solvent used in the production of methamphetamine. It is flammable, but only mildly toxic in the short term (used for drinking alcohol).
- **Ethyl ether**, an ether hydrocarbon derivative, is colorless with a sweet, pungent odor. Ethyl ether is highly flammable, toxic, and is used in the manufacture of methamphetamines. While it is the purpose of illegal drug manufacturers to produce drugs quickly and it is unlikely ethyl ether will be around long enough to become a danger, responders should be aware of the potential of explosive peroxide formation. Where containers of ether are in use for 6 months or longer, these peroxides can form. They are sensitive to heat and shock; any suspected old containers of ether should be treated like potential bombs.
- **Formic acid**, also an organic acid, is a colorless liquid with a pungent odor. It is used in the process of methamphetamine manufacture. Formic acid is corrosive and toxic. Contact with oxidizing agents may cause an explosion.
- **Hexane**, a colorless liquid with a mild characteristic odor, is a solvent used in the production of methamphetamines. It is a central nervous system toxin and is extremely flammable.
- **Hydriodic acid** is used by oil refineries to test crude oil for sulfur content. It is the principal chemical in the pseudoephedrine reduction process. Hydriodic acid breaks down the pseudoephedrine molecules to create methamphetamines.
- **Hydrogen iodide** is a colorless gas used as a reagent with red phosphorus in the manufacture of methamphetamines. It is corrosive and an irritant.
- **Hydrochloric acid** (muriatic acid or pool acid) is a colorless inorganic acid with a pungent odor. It is used in the manufacture of methamphetamines and is corrosive and has toxic irritating fumes. Gases released during the manufacturing process can be flammable and explosive. It is found in hardware stores.

- **Iodine** is an element that is solid purple crystals or flakes with a sharp odor. It is used in the synthesis of hydriodic acid. Iodine is toxic by inhalation and ingestion and corrosive. It is sometimes used by ranchers to treat thrush on horse hooves. It is used in the initial stages of the pseudoephedrine/ephedrine cooking process.
- **Iodine tincture** (solution with alcohol) is a dark red solution with a medicinal odor that is used in the synthesis of hydriodic acid. It is flammable and toxic by inhalation.
- **Lead acetate** is found as solid white crystals or brown or gray lumps that are odorless. It is used in P2P synthesis and is hazardous from chronic exposure.
- **Lithium aluminum hydride** is a solid white-to-gray powder and is odorless. It is corrosive and extremely water reactive. When in contact with water, it will generate explosive hydrogen gas. It is used in the hydrogenation process during methamphetamine production.
- **Methyl alcohol** (methanol) is an alcohol hydrocarbon derivative and a clear colorless liquid with a characteristic odor. It is a solvent used in the production of amphetamines. Methanol is toxic and flammable and can cause blindness.
- **Naphtha** is a reddish-brown liquid with an aromatic odor. It is a petroleum distillate solvent used in the manufacture of methamphetamine. The primary hazard is toxicity.
- **Phenylacetic acid** is an organic acid that is a solid white, shiny crystal with a floral odor. It is used as a precursor to the synthesis of P2P. It is an irritant and a possible teratogen (causes birth defects).
- **Phosgene**, a toxic gas, is a byproduct of the pseudoephedrine reduction process. When red phosphorus and iodine are heated, the lethal and odorless gas called phosgene is created. This could be dangerous to law enforcement and other responders walking in on an active lab.
- **Phosphine** is a colorless gas with a fish- or garlic-like odor. It is a product of methamphetamine production. It is highly flammable and reacts explosively with air and is toxic by inhalation.
- **Phosphoric acid** is an inorganic acid that is hygroscopic (absorbs moisture from the air) colorless crystal. Its primary hazard is as an irritant, and it is corrosive. It is used as a precursor in the production of methamphetamine.
- **Pseudoephedrine**, a white crystalline powder, is a precursor used in the production of methamphetamines. It is an irritant and is toxic by ingestion. Available in over-the-counter decongestants and diet pills, it is used in the pseudoephedrine/ephedrine reduction method. The federal government has a limit of eight packages that can be purchased per person on medications containing pseudoephedrine.

- **Red phosphorus** is a red-to-violet odorless solid. It is used as a catalyst in methamphetamine synthesis. Catalysts are used to control the speed of chemical reactions. Red phosphorus can form phosphine gas during the production process which is toxic by inhalation, flammable, and reacts explosively with air. It can be found at the end of every matchstick in your home and is also used in road flares.
- **Ronsonol** (lighter fluid) is a reddish-brown liquid with an aromatic odor. It is a petroleum distillate solvent consisting of two solvent naphtha fractions, light aliphatic 95% and medium 5%, and Shell Sol RB 100%. Its properties are similar to naphtha.
- **Sodium metal** is an element that is a solid silvery-white odorless metal or crystal. It is used in the hydrogenation in methamphetamine synthesis. Sodium metal is corrosive and extremely water reactive, liberating hydrogen gas.
- **Sodium hydroxide** (lye) is found as white pellets or flakes that are odorless. It is a reagent used in the manufacture of methamphetamine. Sodium hydroxide is extremely corrosive. When in contact with metals such as sodium or fire, it can produce explosive hydrogen gas.
- **Sulfuric acid** (drain cleaner) is a colorless-to-yellow viscous liquid that is generally odorless. It is used in the manufacture of amphetamine and methamphetamines. It is corrosive and may produce corrosive fumes, and is found in battery acid or drain cleaners.
- **Toluene** is a clear colorless liquid member of the aromatic hydrocarbon family with benzene-like odor. It is an irritant and is highly flammable and is a solvent used in manufacture of P2P and methamphetamine. It is found in paint thinners.
- **1,1,2-Trichloro-1,2,2-trifluoroethane** (Freon) is a clear colorless liquid with a slight ethereal (ether like) odor. It is a solvent used to extract methamphetamine. It is an irritant and is toxic by inhalation.
- **White gas** is a solvent used to extract methamphetamine. It is a colorless liquid that is flammable. White gas is used as a fuel for camp stoves and is readily available in hardware, discount, and home improvement stores.

Home-Made Ammonia

Methamphetamine manufacturing is accomplished in several ways, but one that is becoming popular is the use of anhydrous ammonia, sodium or lithium metal, and the over-the-counter cold medications pseudoephedrine or ephedrine. Heat may be used but is not required. Other materials required include coffee filters, solvents, and other common items easily obtained. (This is also known as the "Nazi" method because the Germans used it during World War II when German soldiers were given methamphetamines to allow them to keep going on a limited diet.)

Figure 4.132 Anhydrous ammonia is heavily regulated and difficult to obtain for clandestine use, so drugmakers often resort to theft from commercial facilities or farmers. (Courtesy: Red Willow Western, NE Rural Fire Department.)

Anhydrous ammonia is heavily regulated and difficult to obtain for clandestine use, so drugmakers often resort to theft from commercial facilities or farmers (Figure 4.132). There is, however, another option: liberating anhydrous ammonia from common garden fertilizers ammonium nitrate and ammonium sulfate. Processing the fertilizer using sodium hydroxide and water liberates ammonia gas. Adding heat to the process speeds up the evolution of ammonia. The liberated ammonia gas is captured and condensed to be used for the methamphetamine manufacture.

Making methamphetamines using this method is not without significant hazard. The uncontrolled release of the ammonia gas, which is flammable, into a closed system is hazardous. The risk of a water reaction with sodium or lithium metal creates an ignition source for the flammable ammonia and hydrogen gas liberated from the water. Ammonium nitrate, an oxidizer, coming in contact with organic fuels can be explosive.

Cleanup Concerns

The cleanup of clandestine drug operations following an investigation by law enforcement personnel is not a function of emergency responders. Cleanup should be coordinated with the state department of the environment using recommended cleanup contractors. Because some aspects of the cleanup may prove dangerous for contractors, emergency personnel may want to keep a crew on-site in case of a fire or release requiring emergency actions.

Personnel entering the contaminated area should only do so using appropriate chemical protective clothing for the hazard present. Contamination may extend beyond the location of the actual drug lab.

Ventilation and plumbing systems within hotels, motels and apartment buildings may also be contaminated and require decontamination.

When drug labs are discovered, they become miniature hazardous waste cleanup sites. It is estimated that as much as 7 lb of waste chemicals are produced for every pound of methamphetamine processed. Waste from drug labs is also sometimes dumped clandestinely on roadsides, vacant property, fields, and wooded areas. Materials dumped to dispose of evidence can include propane containers, empty 2-L soda bottles, containers for other materials (such as starter fluid, brake fluid, brake cleaner, lighter fluid, rock salt, acetone, or camping stove fuel), or miscellaneous glass and other containers. Investigators at the scene of a suspected clandestine drug-making operation need chemical protective clothing and respiratory protection. Hazmat teams may be called upon to assist because of the dangers of some of the chemicals involved (*Firehouse Magazine; Hazardous Materials Chemistry for Emergency Responders*).

Pesticides

A pesticide is a chemical or mixture of chemicals used to destroy, prevent, or control any living thing considered to be a pest (Figure 4.133), including insects (insecticides), fungi (fungicides), rodents (rodenticides), or plants (herbicides). The definition of a pesticide from the Federal Insecticide, Fungicide, and Rodenticide Act (FIFRA) is "a chemical or mixture of chemicals or substances used to repel or combat an animal or plant pest. This includes insects and other invertebrate organisms; all vertebrate pests, e.g., rodents, fish, pest birds, snakes, gophers; all plant pests

Figure 4.133 A pesticide is a chemical or a mixture of chemicals used to destroy, prevent, or control any living thing considered a pest.

growing where not wanted, e.g., weeds; and all microorganisms which may or may not produce disease in humans. Household germicides, plant growth regulators, and plant root destroyers are also included."

Farming and commercial landscaping activities involve the use of restricted pesticides to control weeds, insects, and fungus in agricultural crops and lawns and ornamental plants. Restricted use pesticides are those regulated by the EPA and only allowed for commercial or agricultural use by trained and licensed applicators. Restricted use pesticides are usually more toxic, maybe more environmentally sensitive, found in greater quantities in storage, and packaged in larger containers. When a pesticide is classified as restricted, the label will state "Restricted Use Pesticide" in a box at the top of the front panel.

A statement may also be included describing the reason for the restricted use classification. Usually, another statement will describe the category of certified applicator that can purchase and use the product. Restricted-Use pesticides may only be used by applicators certified by the state or by the EPA. Home yard work often involves the use of general pesticides to control weeds, insects, and fungus in lawns and ornamental trees and bushes. Chemicals approved by the EPA for homeowner use do not require any special training or licensing by the homeowner. Container size is usually small and they can be safely used by following label directions.

The EPA estimates that there are 45,000 accidental pesticide poisonings in the United States each year, where more than 1 billion pounds are manufactured annually. Pesticides can be found in manufacturing facilities, commercial warehouses, agricultural chemical warehouses, farm supply stores, nurseries, farms, supermarkets, discount stores, hardware stores, and other retail outlets. They may also be found in many homes, garages, and storage sheds across the country.

More than 1,000 basic chemicals, mixed with other materials, produce about 35,000 pesticide products. However, when an emergency occurs involving different groups of pesticides, chemicals may become mixed that are not normally found together. This mixing of chemicals may provide toxicology and cleanup problems for emergency responders. Great care should be taken during firefighting operations including overhaul when pesticides and other chemicals are involved. If the fire is allowed to burn off, care should be taken to avoid smoke or fumes that evolve.

Runoff can become contaminated with toxic materials and damage firefighting protective clothing, equipment, apparatus, and the environment. If a decision is made to fight a pesticide fire, runoff should be kept to a minimum. If possible route runoff water to a holding area. Hazardous materials teams are called upon to respond to numerous incidents each year involving pesticides. Care should be taken so as not to overreact to a pesticide spill. A pint bottle of a pesticide broken on the display floor of a lawn and garden center does not necessarily require a full-blown hazmat response and a multi-hour operation to effectively mitigate. Keep in mind those pesticides designed for consumer use involve the opening of a container and mixing with water for application by the end-user, many times without a need for extensive protective clothing. On the other hand, larger quantities of restricted use pesticides, or unrestricted pesticides would need to be handled as any other serious chemical spill. The important thing is to evaluate the incident. Do a risk–benefit analysis to determine the level of response necessary to mitigate the incident safely.

Pesticides like many other groups of chemicals can be segregated into families based upon their chemical make-up and characteristics, including toxicity. Common pesticide families are organophosphates, carbamates, chlorophenols, and organochlorines.

Organophosphates are derivatives of phosphoric acid and are acutely toxic, but are not enduring. They are generally known as all insecticides, which contain phosphorus. Organophosphates break down rapidly in the environment and do not accumulate in the tissues. They are generally much more toxic to vertebrates than other classes of insecticides and are therefore associated with more human poisonings than any other pesticide. They are also closely related to some of the most potent nerve agents including sarin and VX. Organophosphates function by overstimulating, and then inhibiting neural transmission, primarily in the nervous, respiratory, and circulatory systems.

Signs and symptoms of exposure include pinpoint pupils, blurred vision, tearing, salivation, and sweating. Pulse rate will decrease and breathing will become labored. Intestines and bladder may evacuate their contents. Muscles will become weak and uncomfortable. Additional symptoms include headache, dizziness, muscle twitching, tremor, or nausea. Symptoms of most pesticide poisonings are similar. Examples of organophosphate pesticides include malathion, methylparathion, thimete, counter, lorisban, and dursban. The chemical formulas of the organophosphates contain carbon, hydrogen, phosphorus, and at least one sulfur atom, and some may contain at least one nitrogen atom. Antidotes are available for organophosphate pesticide poisonings. Many hospitals in rural areas where organophosphates are used by farmers and others will have extra stocks of atropine, which is used to counter the effects of organophosphate pesticides.

Carbamate pesticides are derivatives of carbamic acid and are among the most widely used pesticides in the world. Most are herbicides and fungicides, such as 2,4,D, paraquat, and dicamba. Carbamate pesticides function in the body by inhibiting nerve impulses. Carbamates contain carbon, hydrogen, nitrogen, and sulfur. Examples include furadan, temik, and sevin.

Organochlorine pesticides are known as chlorinated hydrocarbons, chlorinated organics, chlorinated insecticides, and chlorinated synthetics. They contain carbon, hydrogen, and chlorine, and are neurotoxins, which function by overstimulating the central nervous system, particularly the brain. Examples of organochlorine pesticides include aldrin, endrin, hesadrin, thiodane, and chlordane. The best-known organochlorine is DDT, which has been banned for use in the United States because of its tissue accumulation and environmental persistence. Organophosphates do not break down in the environment, which can affect the food chain. Chlorophenols contain carbon, hydrogen, oxygen, and chlorine. They affect the central nervous system, kidneys, and liver when in contact with the body.

Pesticide Labels

Pesticide labels contain valuable information for the emergency responder and medical personnel treating a patient exposed to pesticides. EPA regulates the information required for pesticide labels. Pesticide users are required by law to comply with all the instructions and directions for use in pesticide labeling. Information on pesticide labeling usually is grouped under headings to make it easier to find the information when needed. These headings may include Identifying Information, Restricted-Use Designation, Front-Panel Precautionary Statements, Hazards to Humans and Domestic Animals, Environmental Hazards, Physical or Chemical Hazards, and Directions for Use.

Information on pesticide labels emergency responders may need includes product name, "Signal Word," a statement of practical treatment, EPA Registration Number and Establishment Number, a note to the physician, and a statement of chemical hazards. EPA registration numbers indicate that the pesticide label has been approved by the EPA. The establishment number identifies the facility where the product was made. Other information on the label useful to responders includes active and inert ingredients. "Inert" does not necessarily mean that the ingredients do not pose a danger; it means only that the inert ingredients do not have any action on the pest for which the pesticide was designed.

Many times the inert ingredient is a flammable or combustible liquid. They can also range in toxicity from extremely toxic to relatively nontoxic. Formulation information may also be provided on the label or maybe abbreviated, such as WP for wettable powder, D for dust, or EC for emulsifiable concentrate. The label also contains information about treatment for exposure. This information should be taken to the hospital when someone has been contaminated with a pesticide. Do not, however, take the pesticide container to the hospital! Take the label or write the information down, take a Polaroid picture of the label or use a pesticide label book. (Label books are available from agricultural supply dealers.)

Another good reference source for information about pesticides is the *Sittig's Handbook of Pesticides and Agricultural Chemicals.*

Pesticide Toxicity (Signal Words)

Pesticides can be grouped generally into three toxicity categories: High, Moderate, and Low. Signal words corresponding with the level of toxicity and the caution statement "Keep out of the reach of children" must appear on every label. Three signal words that indicate the level of toxicity of a pesticide include "DANGER," CAUTION," and WARNING." Highly toxic materials bear the word "DANGER." All highly toxic pesticides that are very likely to cause acute illness through oral, dermal, or inhalation exposure also will carry the word POISON printed in red and the skull and crossbones symbol.

Products that have the signal word danger due only to the skin and eye irritation potential will not carry the word POISON or the skull and crossbones symbol. The lethal dose for products marked with DANGER on the label maybe a few drops to 1 tsp. Moderately toxic pesticides have the word "WARNING." This word signals that the product is moderately likely to cause acute illness from oral, dermal, or inhalation exposure or that the product is likely to cause moderate skin or eye irritation. The lethal dose for moderately toxic pesticides is 1 tsp to 1 tbsp. Low toxicity pesticides carry the word "CAUTION." This word signals that the product is slightly toxic or relatively nontoxic. The product has only slight potential to cause acute illness from oral, dermal, or inhalation exposure. The skin or eye irritation it would cause, if any, is likely to be slight. The lethal dose for pesticides with CAUTION on the label is 1 oz to 1 pt.

Toxicity

Pesticides may poison or cause harm to humans by entering the body in one or more of these four ways: (1) through the eyes, (2) through the skin, (3) by inhalation, and (4) by swallowing. As with most any chemical, exposure to the eyes is the fastest way to become poisoned. The route of entry is the method the pesticide enters the body. However, it may not be the point where the injury occurs. For example, a pesticide may enter the body through inhalation, but may have an effect on the central nervous system, and not affect the respiratory system. In this instance, the route of exposure is inhalation through the lungs; the target organ is the central nervous system.

Whenever anyone is exposed to a pesticide, it is important to recognize the signs and symptoms of poisoning, so that prompt medical help can be provided. Any unusual appearance or feeling of discomfort or illness can be a sign or symptom of pesticide poisoning. These signs and symptoms may be delayed up to 12h. When they occur and pesticide contact is suspected, get medical attention immediately. (The National

Pesticide Network, located in Texas, provides emergency information through a toll-free telephone number, 800-858-7378, from 8 a.m. to 6 p.m. Central Standard Time.) Information can also be obtained from the **National Poison Control Center** by calling **1-800-222-1222.**

While emergency responders should never risk life to protect the environment, they certainly shouldn't do anything to damage the environment further than if they had not responded at all. Some pesticide fires may need to be left to burn out on their own to avoid a contaminated runoff, which could damage the environment. Many times fire can break down pesticides into less harmful chemicals when they burn. Pesticide labels contain information about what type of environmental damage can occur from the uncontrolled release of the pesticide. Care should be taken to protect the environment in consultation with state and local environmental officials (Hazardous Materials Chemistry for Emergency Responders).

Plastics & Polymers

Plastics have been around for more than 100 years. The first plastic developed was cellulose nitrate, which was a replacement for ivory in billiard balls. Since World War II, however, the plastics industry has been one of the fastest developing technologies. The forms, variations, and applications of plastics have developed at a tremendous rate, producing a family of materials that is unusually complicated and diverse.

Plastics are composed of organic materials that are part of a group of materials known as polymers. Polymers can be subdivided into two groups – naturally occurring and man-made. Common naturally occurring polymers include leather, wood, paper, silk, cotton, and wool. Man-made polymers are created from organic materials found in nature. Approximately 53 billion pounds of synthetic polymers are manufactured each year in the United States alone.

The American Society for Testing Materials (ASTM) defines plastic as "a material that contains as an essential ingredient one or more organic polymeric substances of large molecular weight, is solid in its finished state, and at some stage in its manufacture or processing into finished articles can be shaped by flow."

Terminology

The word polymer comes from the Greek "poly" (many) and "mer" (part). Therefore, a polymer is a compound with many parts. A polymer is a long-chained molecule composed of many smaller parts called monomers ("mono" is Greek for one). Monomers, however, are able to hook together into long chains of hundreds or thousands of parts. Such hooking is called polymerization. In polymerization, monomers – which have double bonds – are broken down by heat or chemical reaction to single bonds. The single bonds attach to each other creating a self-reaction, which in turn creates the long-chained polymer. This reaction usually takes place in a reactor vessel in a chemical plant under controlled conditions. But it can also take place during transportation or in storage, creating a hazard for emergency responders.

Man-made and naturally occurring polymers behave much the same way during reactions, especially when exposed to fire. In fact, some synthetic polymers contain the same elements as natural polymers, exhibit the same burning characteristics, and produce the same products of combustion – many of which can be toxic. Naturally occurring polymers are standard in their identities – wood, for example, is not mistaken for cotton or wool. Synthetic polymers, however, are produced to conform to specific properties, and it may be difficult to distinguish between them. For example, polypropylene and polyethylene are very similar, as are styrene and styrene-acrylonitrile.

Many terms are interchangeable when referring to synthetic polymers. The terms plastic, polymer, resin, compound, and high-polymer macromolecular substance are used often used interchangeably.

Plastics are such a common part of our everyday lives that we don't give them much thought. They are found everywhere. Look around the room you are in, how many things are made of natural polymers and how many are synthetic?

As I type this volume on my computer, I see that the housings for the computer as well as the keyboard, printer, and monitor are made of plastic; the table the computer sits on is plastic; the chair I am sitting on is largely made of plastic and plastic fibers; the carpeting on the floor is made from plastic fibers; the telephone on the table is plastic; the cabinets next to the computer look like wood, but are actually plastic; the list could go on and on.

The jobs of emergency responders are made easier and safer because of plastics. Nearly all components of firefighter turnouts are made of plastic and plastic fibers; chemical suits, gloves, and boots worn by hazmat technicians are made of plastic; ropes we use for rescue are made of plastic fibers; and many EMS supplies are either plastic or packaged in plastic. Again, the list could go on and on. The polymer family tree has three distinct branches or divisions, and each branch may be identified by particular characteristics. Those branches are thermosets, thermoplastics, and elastomers.

Thermosets are plastics that are hardened into their final shape in the manufacturing process by heating and usually cannot be softened again by heating without losing their identity. If attempts are made to re-form them by heat, they will start to burn. Those that can be softened by heating cannot be remelted or returned to their original form before they harden once again.

Common Examples of Polymers

Thermosets	Thermoplastics	Elastomers
Epoxies	ABS (acrylonitrile, butadiene, and styrene)	Ethylene
Butadiene	Polyesters	Nylons
Natural Rubber	Polyurethane	Vinyls
Neoprene	Silicones	Acrylics

Thermoplastics are compounds formed by heat and pressure, then cooled into their final shape. They can be re-formed again if placed under heat or pressure, or both. In some cases, they can be heated and reformed up to 20 times without losing their properties.

Elastomers are sometimes called "synthetic rubbers" because of their elastic or rubber-like features. Natural rubber is also classified as a polymer that would be in the elastomer branch of the family tree. Synthetic rubbers were created to duplicate or surpass the most desirable properties of natural rubber. Elastomers are substances that at room temperature can stretch to at least twice, and in some cases many times, their original size and upon release return to their previous form with notable force. Elastomers cannot be heated and re-formed.

Manufacturing Plastics

The process of making plastics involves many different compounds and hazardous materials that are commonly shipped in transportation and stored in manufacturing facilities. One of the most common compounds used in the making of plastics is the monomer. Monomers can be found as solids, liquids, or gases.

Chemical Notebook

Ethylene, which is used to make polyethylene plastic, is a colorless gas with a sweet odor and taste. It is also a highly flammable gas with a wide flammable range of 3%–36% in air. It is not considered toxic but can displace oxygen in the air and create an asphyxiation hazard to response personnel. (Ethylene is also the gas that is produced naturally by ripening fruit and is used in orchards to hasten fruit ripening.)

Styrene is another monomer from the aromatic hydrocarbon family, along with benzene, toluene, and xylene. Also called vinylbenzene, it is a colorless, oily, aromatic liquid. Styrene is a moderate fire risk with a narrow flammable range of 1.1%–6.1% in air. It is toxic by ingestion and inhalation and has a threshold limit value (TLV) of 50 parts per million (ppm) in air. Styrene monomer is used to make polystyrene, which is the rigid plastic from which soft-drink-cup covers are made. Polyacrylamide, a solid monomer white in color, is used to make the clear plastic from which compact discs are made.

Butadiene is another common, but dangerous monomer. It is a highly flammable gas with a flammable range of 2%–11% in air. Butadiene is also a suspected carcinogen and has a TLV of 10 ppm in air. It is used in making elastomers and neoprene.

Nylon is a polymer having a high molecular weight. It is formed from adipic acid and hexamethylenediamine, which produces a material with high strength, elasticity, abrasion resistance, and solvent resistance. Nylon is used to make tire cords, rope, and apparel, and has other industrial and automotive uses.

Kevlar is a form of nylon that is used in some firefighter gloves and other safety products because of its great strength and resistance to puncture. Kevlar also has high thermal endurance and doesn't start melting until it tops 950°F.

Monomers cannot be shipped or stored unless they have been stabilized through the use of an inhibitor. The inhibitor prevents the uncontrolled polymerization of the monomer during transportation and storage. It does not interact chemically with the monomer, it just acts as a stabilizer. U.S. DOT regulations prohibit the transportation of most monomers without the material being inhibited.

Author's Note: *One of the changes to the 1996 Emergency Response Guide Book was the identification of materials in transportation that may undergo polymerization creating a danger to response personnel. When looking up a chemical by its four-digit identification number in the yellow section or alphabetically in the blue section of the book, you will be referred to a three-digit guide page in the orange section of the book. That number will have a "P" after it if the material in question has the potential to undergo polymerization.*

You may also notice that next to the common shipping name of the material in the ERG will appear the words "inhibited" or "stabilized." During an accident involving a railcar or tanker truck containing a monomer, the inhibitor can be separated from the monomer, enabling the monomer to undergo an explosive polymerization reaction that could cause the container to fail violently.

Once monomers and other chemicals are processed into plastic materials, they generally do not present any danger to civilians or emergency responders. Many plastic items, in fact, are used to sustain and improve the quality of life – artificial hearts, replacement joints, medical tubing, and packaging.

The greatest concern for emergency responders, and particularly firefighters, is the burning of plastics. All plastics that contain carbon will burn. As with any materials that will burn, some are more combustible than others. Chemicals can be combined with plastics during manufacture to reduce combustibility. Other plastic materials are formulated to be self-extinguishing. Plastics that contain carbon and hydrogen – such as polyethylene, polypropylene, polybutylene, and polystyrene – burn very well. Burning polyethylene melts smells like wax and produces dripping of the melted flaming material that could spread the fire. Polystyrene burns much slower, producing large soot particles, and smells like vinegar. Styrene is an aromatic hydrocarbon and they, as a family, burn sooty with incomplete combustion.

Plastics that are composed of carbon, hydrogen, and oxygen burn slower than the others mentioned. Thermosetting plastics can produce burning smells like charred wood or formaldehyde. Those plastics that contain nitrogen and sulfur will produce very toxic gases when they burn. Plastics containing nitrogen burn with no smell, and those with sulfur produce a choking sulfur dioxide smell.

Those plastics that contain halogens (fluorine and chlorine) produce acrid, choking odors when they burn. With fluorine- and chlorine-based plastics, the flame must be continuous or the plastic will self-extinguish. Almost all plastics can be made flame or smoke retardant by adding other chemicals during manufacturing. Adding any of the halogens will retard burning. Halogens are used with carbon to produce halon fire-extinguishing agents. Silicone-based plastics will not burn at all.

Flammable Products

When plastics burn, the rate of burn and amount of smoke generated can vary widely depending on the type of plastic and flame retardant chemicals present. Cellulose nitrate motion picture film was used for many years in Hollywood. Cellulose nitrate is highly flammable and ignites readily. Cellulose nitrate is made by reacting sulfuric acid and nitric acid on cellulose materials such as cotton. Cellulose nitrate plastics exhibit dangerous burning characteristics, unlike any other plastic. When exposed to heat, the physical make-up of cellulose nitrate is changed in such a way that it may become subject to spontaneous combustion.

Combustion products produced when cellulose nitrate burns can be very toxic. Use of cellulose nitrate is not as widespread as it once was, but if encountered should be handled very carefully. New safer plastics have been developed to replace cellulose nitrates in such products as photographic and motion picture film and housewares. Teflon, on the other hand, another popular plastic, is not combustible at all and experiences heat damage only at extremely high temperatures.

Some plastics may exhibit unusual burning characteristics compared to building materials made from natural polymers such as wood. Plastics as a group generally have higher ignition temperatures than wood and other cellulose building products.

Plastics have been reported to have very high flame spread characteristics, as high as two feet per second, or 10 times that of wood on the surface. Vinyl, when tested in a solid form in the laboratory, has been shown to burn slowly. However, when in the form of a thin coating on wall coverings, it spreads rapidly contributing to flame spread.

Nylon has a tendency to self extinguish when a flame is removed. When nylon is in the form of carpet fiber, under certain conditions it burns with great enthusiasm. Polyurethane foam that has not been treated with a flame retardant is very flammable. It is used as an insulating material in construction and burns with a very smoky flame. Because of its burning characteristics, it has contributed to rapid flame spread in several fatal fires.

Burning of plastics may produce large quantities of thick, black smoke. When chemicals have been added to retard burning, they may actually expand the amount of smoke that is produced. Because of their ability to melt and run, plastics can spread fires in ways that could mislead fire investigators. When skylights and light fixture diffusers are ignited by high ceiling temperatures, they may soften and sag. They can fall into combustible materials below and start fires in several isolated locations. This could lead an investigator to suspect an incendiary fire when in fact it was not.

Dangers to Firefighters

The products of combustion when plastics and other materials burn is the most significant hazard to both occupants of a building and firefighters during a fire. Plastic materials that contain only carbon and hydrogen will generate only carbon, carbon monoxide, carbon dioxide, and water as they burn.

Intermediate products of combustion, however, also are produced and can include acrolein, formaldehyde, acetaldehyde, propionaldehyde, and butyraldehyde. Members of the aldehyde hydrocarbon family are irritants and flammable, with wide flammable ranges. Acrolein and carbon monoxide are lethal poisons. In addition to being toxic, carbon monoxide is also extremely flammable.

Examples of carbon- and hydrogen-based plastics include polyethylene, polypropylene, and polystyrene. Combustion products produced from burning plastics containing only carbon and hydrogen are the same as natural polymers such as wood, paper, and other Class A combustible materials. Plastics containing carbon, hydrogen, and oxygen produce the same combustion products as those containing just carbon and hydrogen. The major difference is these plastics tend as a group to be less combustible than those with just carbon and hydrogen. Plastics in this group include acetal, acrylics, alyls, cellulosics, some epoxies, ethylene vinyl acetate, ionomers, phenolics, polycarbonate, polyesters, and polyphenylene oxide. When nitrogen is added to a plastic compound, an additional product of combustion occurs when it burns.

Another product of combustion is hydrogen cyanide, an extremely toxic and flammable gas. Just like carbon monoxide, much of the hydrogen cyanide is consumed by the fire as it is produced. (It must be noted that hydrogen cyanide is also generated by the burning of the natural polymers including, leather, wood, silk, and some types of paper.)

Plastics that contain fluorine and chlorine are generally less combustible than other plastics. As they decompose when exposed to the heat of a fire, they produce hydrogen fluoride and hydrogen chloride gases, respectively. Both are corrosive gases that can form acid when in contact with moisture, such as in the eyes, skin or lungs.

Plastics containing fluorine include fluoroplastics, of which Teflon (polytetrafluoroethylene) is the most common. Chlorine containing plastics include polyvinyl chloride (PVC), polyvinyl dichloride, and polyvinylidene chloride. Plastic compounds can also contain sulfur, which will decompose to sulfur dioxide during the combustion process. It is a strong irritant and can be lethal when exposed to concentrations of 100 ppm for more than 10 min. Plastics that contain sulfur include polysulfone and polyethersulfone.

Firefighters who wear SCBA will be protected from carbon monoxide and other toxic fire gases. Many of these gases are invisible and remain present even after the smoke is "cleared." Firefighters should continue to wear SCBA during overhaul operations and not remove them until they are outside the fire building or until the air has been monitored within the building and shown to be safe.

It is also important to remember that the same toxic materials inside the fire and smoke can become impregnated in firefighter turnouts. Each time the unwashed turnouts are worn, the firefighter is re-exposed to the toxic substances in the turnouts. Firefighter turnouts should be washed or decontaminated to help prevent this continued exposure to toxic materials (*Firehouse Magazine; Hazardous Materials Chemistry for Emergency Responders*).

Spontaneous Combustion

Spontaneous combustion is a sometimes mysterious and often misunderstood or misdiagnosed combustion of flammable materials. I am sure that many fires have been blamed on spontaneous combustion that were not and many fires that were caused by spontaneous combustion were not properly attributed. Certainly, fire investigators and hazardous materials personnel – if not all firefighters – should know the concept of spontaneous combustion, the types of materials that may undergo such a reaction, and the conditions necessary for it to happen.

Spontaneous combustion, according to the *Handbook of Fire Prevention Engineering*, "is a runaway temperature rise in a body of combustible material that results from the heat being generated by some process taking place within the body." Spontaneous combustion may be rapid or slow. It can result from reactions of a susceptible material with air or water or from a chemical reaction. Materials involved can be chemicals, elements, or hydrocarbon compounds or a mixture.

Slow spontaneous combustion can occur in two general ways: biological processes of some microorganisms and slow oxidation. Biological processes occur within organic materials such as hay and grass clippings. The activity of biological organisms within the material generates heat that is confined by the materials themselves until the ignition temperature is reached and ignition occurs.

Slow oxidation is a chemical reaction. Chemical reactions may produce heat; reactions that produce heat are considered exothermic. If the heat is insulated from dissipating to the outside of the material, it will continue to build up. As the heat builds, the material is heated from within. The process continues until the ignition temperature of the material is reached and ignition occurs.

Hydrocarbon compounds usually undergo slow spontaneous combustion. Generally, hydrocarbon compounds are considered saturated or unsaturated. Within saturated compounds, all of the chemical bonds among the elements in the compound are single. Single bonds are "full," for there isn't room or a method for another element to attach. Double or triple bonds are unsaturated. In other words, the double bond can break, and when it does, another element can attach. Spontaneous ignition occurs when this double or triple bond breaks and creates heat and the heat is confined. The material itself produces sufficient heat to reach its own ignition temperature.

Some materials subject to spontaneous combustion are not considered hazardous in transit by the U.S. DOT and may not be placarded or labeled. Materials may have other hazards considered more severe. DOT has assigned a hazard class to materials that are shipped in transportation and are subject to spontaneous combustion by chemical nature. Flammable solids, Division 4.2 materials are spontaneously combustible. The DOT defines them as pyrophoric materials.

Even though this hazard class is flammable solids, these materials may be found as solids or liquids. They can ignite without an external ignition source within 5 min after coming into contact with air. There are other 4.2 materials that may be self-heating, i.e., in contact with air and without an energy supply (ignition source); they are liable to self-heat, which can result in a fire involving the material or other combustibles nearby. This type of spontaneous combustion is considered rapid.

Materials can also spontaneously combust when in contact with water. When some carbon-based materials, such as activated carbon or charcoal briquettes, are in contact with water, an oxidation reaction occurs between the carbon material, the water, and pockets of trapped air. This type of spontaneous combustion occurs slowly. The reaction is exothermic, which means heat is produced in the reaction and slowly builds up until ignition occurs spontaneously. Because the reaction is so slow, we use charcoal lighter rather than water when we are cooking on the grill!

Materials subject to spontaneous heating are listed below:

Alfalfa meal	Animal hides	Castor oil	Coal
Cottonseed oil	Fertilizers	Fish meal	Lanolin
Lard oil	Linseed oil	Manure	Metal powders
Olive oil	Peanut oil	Powdered eggs	Soybean oil
Used burlap	Whale oil		

Hidden Hazard

Some types of combustible liquids, such as animal and vegetable oils, have a hidden hazard: they may burn spontaneously when improperly handled. They have high boiling and flashpoints, narrow flammable ranges and low ignition temperatures, and are nonpolar. Carbon-based animal or vegetable oils, such as linseed oil, cooking oil, cottonseed oil, corn oil, soybean oil, lard, and margarine, can undergo spontaneous combustion when in contact with rags, cardboard, paper, or other combustibles. These unsaturated compounds can be dangerous when combustible materials containing residue are not properly disposed of or they come in contact with other combustible materials.

Susceptible Double Bonds

There is a double bond in the chemical make-up of animal and vegetable oils that react with oxygen in the air. The oxygen from the air trapped in the mass reacts with the double bonds present in the animal and vegetable oils. The breaking of the double bonds creates heat. If the heat is allowed to build up in a pile of rags, spontaneous combustion will occur over a period of hours. Fires started by this spontaneous heating process can be difficult to extinguish because they usually involve deep-seated fires. For enough heat to be sustained to cause combustion, there must be insulation. This insulation can be the material itself or maybe in the form of some other combustible material.

Chemicals that Do Not Spontaneously Combust

Spontaneous heating cannot occur in the case of petroleum oils or other hydrocarbon materials that are saturated. Ordinary petroleum products, such as motor oil, grease, diesel fuel, and gasoline, do not have a double bond in their chemical make-up. For that reason, the oxidation reaction that occurs with animal and vegetable oils and the oxygen in the air does not occur. Therefore, those materials do not undergo spontaneous combustion! This fact may come as a surprise to some people because there have been numerous fires blamed on soiled rags with those products on them. The fact is that saturated flammable liquids do not spontaneously ignite and cannot start to burn without some other ignition source.

Laundry Fires

A series of fires have occurred in laundries around the country since 1989. One in every six commercial, industrial, or institutional laundries reports a fire each year, which results in over 3,000 fires. The primary cause is thought to be spontaneous combustion. Chemicals, including animal and vegetable oils, may be left behind in fabrics after laundering. The heat from drying may cause the initiation of the chemical reaction that causes spontaneous ignition.

On June 16, 1992, a fire in a nursing home laundry in Litchfield, IL, caused $1.5 million in damage. The cause was determined to be spontaneous ignition of residual chemicals in the laundered fabric reacting to heat from the dryer. In Findlay, OH, on July 2, 1994, a fire destroyed a commercial laundry and caused over $5 million in damage. Traces of linseed oil were found in a pile of clean, warm garments in a cart.

Restaurant Fires

Fires in restaurants have also occurred involving residual animal or vegetable oils in cleaning rags. The oils are never completely removed by laundering. When placed in the dryer, the rags are heated. When they are put away on a storage shelf, this heat can become trapped, along with the oil remaining on rags, when confined. The spontaneous combustion process begins very slowly and the heat of the reaction increases until combustion occurs (*Firehouse Magazine*).

CASE STUDIES

VERDIGRIS, OK

In Verdigris, OK, a fire occurred in an aircraft hangar at a small airport. The owner's living quarters were on the second level of the hangar. Workers had been polishing wooden parts of an airplane in the afternoon. The rags used to apply linseed oil were placed in a plastic container in a storage room in the hangar, just below the living quarters. At around 2 a.m., the rags with the linseed oil spontaneously ignited and the fire traveled up the wall into the living quarters. Fortunately, the owner had smoke detectors; the family was awakened and the fire department was called promptly. The fire was quickly extinguished with a minimum of damage. The V-pattern on the wall led right back to the box where the linseed oil-soaked rags had been placed. There was little doubt what had happened; the confinement of the pile allowed the heat to build up as the double bonds were broken in the linseed oil, which combined with oxygen in the air and spontaneous combustion occurred.

ONE MERIDIAN PLAZA HIGH RISE FIRE

At about 8 p.m. on Saturday, 23 February 1991, linseed oil-soaked rags left behind by a cleaning crew burst into flames on the 22nd floor of the 38-story One Meridian Plaza in downtown Philadelphia (Figure 4.134). The fire quickly spread, unimpeded by fire sprinklers, throughout the 22nd floor and then upward. Sprinklers were not required by the City's building code at the time of construction and were being added to the building only as the opportunity presented itself.

The 12-alarm fire burned for 18h. The extreme heat caused window glass and frames to melt and concrete floor slabs and steel beams to buckle and sag dramatically. Large shards of window glass fell from the facade, cutting through fire hoses on the ground around the building. Three firefighters were trapped on a fully engulfed floor, and efforts to rescue them failed.

The fire would not yield and there were increasing concerns about the stability of the structure. Fire officials called off the attack and allowed the fire to "free burn," concentrating their efforts on containing the fire to this building. When the fire reached the 30th floor, a tenant-installed fire-sprinkler system was activated, and the worst high-rise fire in U.S. history was finally brought under control. The fire caused three firefighter fatalities (LODD) and injuries to 24 firefighters (Philadelphia Fire Department).

LINCOLN, NE

On August 11, 2003, a fire occurred at an auto shop in Lincoln, NE, when a van inside caught fire. Investigation indicated that the cause of the fire was spontaneous combustion from failure to properly dispose of linseed oil-soaked rags (*Firehouse Magazine*).

Some common materials that by chemical nature are subject to spontaneous combustion are presented here for informational purposes. During an incident involving these or any other hazardous materials, the compounds should be looked up in reference materials to determine hazards and proper tactics.

Figure 4.134 At about 8 p.m. on Saturday, 23 February 1991, linseed oil-soaked rags left behind by a cleaning crew burst into flames on the 22nd floor of the 38-story One Meridian Plaza in downtown Philadelphia. (Courtesy: Philadelphia Fire Department.)

Chemical Notebook

Diethyl zinc. Diethyl zinc is an organo-metal compound and a dangerous fire hazard. It spontaneously ignites in air and reacts violently with water, releasing flammable vapors and heat. It is a colorless pyrophoric liquid with a specific gravity of 1.2, which is heavier than water, so it will sink to the bottom. It decomposes explosively at 248°F. It has a boiling point of 243°F, a flashpoint of –20°F, and a melting point of –18°F. The 4-digit identification number is 1366. The NFPA 704 designation is Health 3, Flammability 4, and Reactivity 3. The white space at the bottom of the diamond has a W with a slash through it to indicate water reactivity. The primary uses of diethyl zinc are in the polymerization of olefins, high-energy aircraft and missile fuel, and the production of ethyl mercuric chloride.

Pentaborane. Pentaborane is a nonmetallic, colorless liquid with a pungent odor. It decomposes at 300°F if it has not already ignited and will ignite spontaneously in the air if impure. It is a dangerous fire and explosion risk, with a flammable range of 0.46%–98% in air. The boiling point is 145°F, the flashpoint is 86°F and the ignition temperature is 95°F, which is extremely low. Any object that is 95°F or above can be an ignition source. Ignition sources can be ordinary objects on a hot day in the summer, such as the pavement, metal on vehicles, and even the air. The 4-digit identification number is 1380. The NFPA 704 designation for pentaborane is Health 4, Flammability 4, and Reactivity 2. The primary uses are as fuel for air-breathing engines and as a propellant.

Aluminum alkyls. Aluminum alkyls are colorless liquids or solids. They are pyrophoric and may ignite spontaneously in air. Aluminum alkyls are pyrophoric materials in a flammable solvent. The vapors are heavier than air, water-reactive, and corrosive. Decomposition begins at 350°F. The 4-digit identification number is 3051. The NFPA 704 designation is Health 3, Flammability 4, and Reactivity 3. The white space at the bottom of the diamond has a W with a slash through it, indicating water reactivity. They are used as catalysts in polymerization reactions.

Aluminum phosphide. Aluminum phosphide (AlP) is a binary salt. These salts have the specific hazard of giving off poisonous and pyrophoric phosphine gas when in contact with moist air, water, or steam. They will also ignite spontaneously on contact with air. This compound is composed of gray or dark yellow crystals and is a dangerous fire risk. Aluminum phosphide decomposes on contact with water and has a specific gravity of 2.85, which is heavier than water. The 4-digit identification number is 1397. The NFPA 704 designation is Health 4, Flammability 4, and Reactivity 2. The white section at the bottom of the diamond has a W with a slash through it, indicating water reactivity. Aluminum phosphide is used in insecticides, fumigants, and semiconductor technology.

Potassium sulfide. Potassium sulfide (K_2S) is a binary salt. It is a red or yellow-red crystalline mass or fused solid. It is deliquescent in air, which means it absorbs water from the air, and is also soluble in water. Potassium sulfide is a dangerous fire risk and may ignite spontaneously. It is explosive in the form of dust and powder. It decomposes at 1,562°F and melts at 1,674°F. Its specific gravity is 1.74, which is heavier than air. The 4-digit identification number is 1382. The NFPA 704 designation is Health 3, Flammability 1, and Reactivity 0. Potassium sulfide is used primarily in analytical chemistry and medicine.

Sodium hydride. Sodium hydride (NaH) is a binary salt that has a specific hazard of releasing hydrogen in contact with water. It is an odorless powder that is violently water-reactive. The 4-digit identification number is 1427. The NFPA 704 designation is Health 3, Flammability 3, and Reactivity 2. The white space at the bottom of the diamond has a W with a slash through it, indicating water reactivity.

White phosphorus (P), also known as yellow phosphorus, is a nonmetallic element that is found in the form of crystals or a wax-like transparent solid. It ignites spontaneously in air at 86°F, which is also its ignition temperature. White phosphorus should be stored and shipped under water and away from heat. It is a dangerous fire risk, with a boiling point of 536°F and a melting point of 111°F. The 4-digit identification number is 2447. The NFPA 704 designation is Health 4, Flammability 4, and Reactivity 2. The primary uses are in rodenticides, smoke screens, and analytical chemistry (Hazardous Materials Chemistry for Emergency Responders).

CASE STUDIES

An incident occurred in Gettysburg, PA, involving phosphorus being shipped under water in 55-gallon drums. One drum developed a leak and the water drained off. This allowed the phosphorus to be exposed to air, causing it to spontaneously ignite. The fire spread to the other areas and eventually consumed the entire truck. The ensuing fire was fought with large volumes of water and in the final stages covered with wet sand. Cleanup created problems because, as the phosphorus and sand mixture was shoveled into over-pack drums, the phosphorus was again exposed to air and reignited small fires (*Firehouse Magazine*).

BROWNSON, NE

A train derailment in Brownson, NE resulted in a tank car of phosphorus overturning and the phosphorus igniting upon contact with air (Figure 4.135). Phosphorus is shipped under water so there was water inside the tank car. CHEMTREC was called and responders were

Figure 4.135 A train derailment in Brownson, NE resulted in a tank car of phosphorus overturning and the phosphorus igniting upon contact with air. (Courtesy: Sidney, NE Fire Department.)

told correctly that the phosphorous would not explode. However, the water inside the tank car was turned to steam from the heat of the phosphorus fire. The pressure from the steam caused a boiler-type explosion that had nothing chemically to do with the phosphorus! (*Firehouse Magazine*).

This is just another example of the hidden hazards that emergency responders must be aware of when dealing with hazardous materials. Not only do the hazardous materials have to be considered but also the container and any "inert" materials that may be involved.

Dumpster Fires

Dumpster fires can be one of the most dangerous fires to unsuspecting firefighters. We sometimes take them for granted. Clandestine dumping is a big hidden problem. No type of fire should be taken lightly, but dumpster fires can be deadly. SCBA should always be worn for dumpster fires. Any chemical reactions or minor explosions should cause firefighters to withdraw and let the fire burn itself out. Do not take chances.

CASE STUDY

NIOSH FIREFIGHTER FATALITY INVESTIGATION

In December 2009, a 33-year-old male firefighter died and eight firefighters, including a lieutenant and a junior firefighter, were injured in a dumpster explosion at a foundry in St. Anna, Wisconsin. At 19:33 hours, dispatch reported a dumpster fire at a foundry in a rural area (Figure 4.136). Eight minutes later, the initial responding crews and the incident commander (IC) arrived on scene to find a dumpster emitting approximately 2-ft high bluish-green flames from the open top and having a 10-in. reddish-orange glow in the middle of the dumpster's south side near the bottom. The IC used an attic ladder to examine the contents of the dumpster: aluminum shavings, foundry floor sweepings, and a 55-gallon drum. Approximately 700 gallons of water was put on the fire with no effect.

Approximately 100 gallons of foam solution, starting at 1% and increased to 3%, was then put on the fire, and again there was no

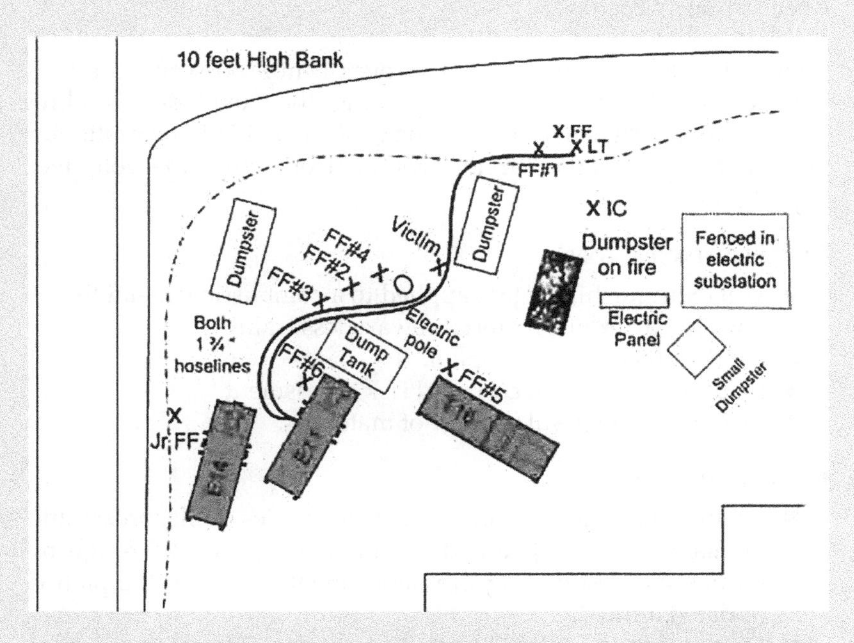

Figure 4.136 Firefighter killed and some others injured in a dumpster explosion at a foundry in Wisconsin. (From Chemical Safety Board.)

Figure 4.137 Approximately 100 gallons of foam solution, starting at 1% and increased to 3%, was put on the fire, with no noticeable effect. (From Chemical Safety Board.)

noticeable effect (Figure 4.137). Just over 12 min on scene, the contents of the dumpster started sparking then exploded sending shrapnel and barrels into the air. The explosion killed one firefighter and injured eight other firefighters, all from the same volunteer department.

Contributing Factors

- Wet extinguishing agent applied to a combustible metal fire.
- Lack of hazardous materials awareness training.
- No documented site preplan.
- Insufficient scene size-up and risk assessment.
- Inadequate disposal/storage of materials.

Key Recommendations

- Ensure that high-risk sites such as foundries, mills, processing plants, etc. are preplanned by conducting a walkthrough by all possible responding fire departments and that the plan is updated annually.
- Ensure that specialized training is acquired for high-risk sites with unique hazards, such as combustible metals.
- Ensure that standard operating guidelines are developed, implemented and enforced.

- Ensure a proper scene size-up and risk assessment when responding to high-risk occupancies such as foundries, mills, processing plants, etc.
- Ensure a documented junior firefighter program that addresses junior firefighters being outside the hazard zone.

Additionally, manufacturing facilities that use combustible metals should:

- Implement measures such as a limited access disposal site and container labeling to control risks to emergency responders from waste fires.
- Implement a bulk dry extinguishing agent storage and delivery system for the fire department.
- Establish a specially trained fire brigade.

Fire Department

The volunteer department involved in this incident had 1 station with approximately 25 volunteer firefighters and 7 fire apparatus serving a population of about 1,000 residents in a geographic area of approximately 25 miles^2. The fire department averages 20 total calls per year. The fire department had been called to this foundry a few times in the past for the smell of smoke in the building, but there were never any fires. The previous calls ended up being overheated drive belts on electrical motors. The fire department had no documented standard operating guidelines (SOGs).

Arrival Timeline of Apparatus and Personnel The fire department is located approximately 5 miles from the foundry.

- 19:33 Hours
  Dispatch reported a dumpster fire at a foundry in a rural area.
- 19:41 Hours
  Engine #11 (E11) – Chief (Driver and IC), captain, a firefighter, and FF #1 (injured)
  Tanker #17 (T17) – 1st Assistant Chief and FF #2 (injured).
- 19:42 Hours
  Engine #14 (E14) – Lieutenant (Lt) (injured), victim, FF #3 (injured), and FF #4 (injured)
  Equipment Truck #15 (ET15) – 2nd Assistant Chief, driver, and FF#6 (injured)
- 19:45 Hours
  Tanker #16 (T16) – FF #5 (injured)
  Personally owned vehicle (POV) – Jr. FF (injured)

Contributing Factors Occupational injuries and fatalities are often the result of one or more contributing factors or key events in a larger sequence of events that ultimately result in the injuries or fatality. The NIOSH investigator identified the following items as key contributing factors in this incident that ultimately led to the line of duty death of one firefighter and to the injuries of eight firefighters:

- Wet extinguishing agent applied to a combustible metal fire.
- Lack of hazardous materials awareness training.
- No documented site preplan.
- Insufficient scene size-up and risk assessment
- Inadequate disposal/storage of materials.

Cause of Death/Injuries According to the medical examiner's autopsy report, the victim died from multiple injuries as a result of blunt force trauma. According to medical records, all of the firefighters' injuries were due to the explosion resulting in debris impact and/or noise-related injuries. Of the firefighters injured at the time of the incident, the Lt had lower back and spinal cord injuries; FF #1, FF#5, FF#6, and the Jr. FF experienced temporary hearing loss; FF#2 had back pain and temporary hearing loss; FF#3 had a neck sprain and a bump on his head; FF#4 had second degree burns to the left elbow, right flank, and a broken right hand (NIOSH).

Chemical Laboratories

There are likely to be few places where firefighters and other emergency response personnel will encounter a wider variety of dangerous chemicals than in a chemical laboratory (Figure 4.138). Laboratories can be found in a variety of locations, including industrial, research facilities, high school, and college labs to name a few. High school and college laboratories are particularly dangerous because the variety of chemicals used and stored there may be greater than any other lab. While the quantities of individual chemicals are usually relatively small, collectively, and in some cases individually, the danger to response personnel can be significant.

Chemicals can be toxic in very small amounts when absorbed through the skin, inhaled, or ingested. They can also cause eye damage, skin burns, illness, and cancer. Every category of the DOT Hazard Classes can be identified among the chemicals in a high school or college laboratory setting. Other labs are more specialized and the range of chemicals is limited to research projects or analytical needs of the facility. When stored with caution, most chemicals do not pose an unreasonable threat under normal conditions. However, chemicals in laboratories are often stored in

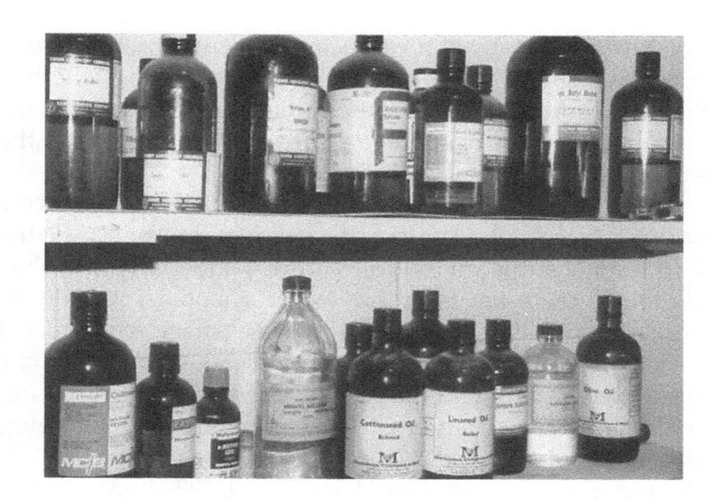

Figure 4.138 There are likely to be few places where firefighters and other emergency response personnel will encounter a wider variety of dangerous chemicals than that in a chemical laboratory.

improper locations and alphabetical order, which may place totally incompatible chemicals on the same shelf with each other. Storing chemicals alphabetically can place dangerous materials such as nitric acid, which is a strong oxidizer, on the same shelf with flammable liquids. Mixing the two can result in an explosive compound. A more appropriate storage system might be one in which organic and inorganic chemical families are stored together. Many chemical supply companies include proper storage system information in their chemical catalogs. Fisher Scientific Company has a very good one in their catalog.

Material Safety Data Sheets (MSDS) also may contain information about chemical compatibilities. MSDS sheets should be maintained on file for each chemical in the laboratory. Manufacturers or chemical suppliers can provide the MSDS sheets upon request. In many cases, they will be shipped automatically with the chemicals. Chemicals that require refrigeration and storage in flammable liquid or acid cabinets are often stored on open shelves in the lab or classroom. Acids and bases, while categorized by the DOT as corrosives, react violently with each other, and should be stored in separate locations and cabinets. Other chemicals can be dangerous under normal conditions when exposed to heat, shock, friction, water, or air. Care must be taken to store these chemicals in safe locations. To make ordinary storage conditions worse, chemicals can degrade, dehydrate, or form other dangerous compounds as they age. Many compounds that are normally safe can be turned into shock or heat-sensitive explosive materials as they age. Many of these factors can create risk for emergency response personnel if they are unaware of the dangers. Oftentimes laboratory personnel and teachers are unfamiliar with the hazards of aging chemicals. Fire inspectors need to be aware of these dangers so they can point them out during laboratory inspections.

In many industrial and educational occupancies, laboratory operations take up only a small portion of the space in the facility. Plating labs are used to maintain the quality of plating solutions within the plant. Chemicals found there are fairly limited and include cyanide, sodium hydroxide, and many different types of acids. Contact between acids and cyanide can produce deadly hydrogen cyanide gas. Other types of industrial labs are used to maintain quality control of manufactured products and chemicals may vary from one facility to the next. Preplanning is an important step in determining the hazards of these facilities.

NFPA Standard 45, as well as NFPA 30, 49, 325, 491M, and 704, are good sources of information concerning fire code issues in laboratory occupancies. Research facilities located in industrial plants or universities also use limited, but very specific chemicals based upon their research. Some of these facilities also have an additional biological hazard over and above the chemical hazards present. Once again, preplanning should ease the minds of response personnel when fires or other emergencies occur at these locations.

Marking systems such as NFPA 704, Hazard Communication, or GIS should be used to identify areas within buildings where dangerous chemical or biological materials may be used and stored along with emergency contact information for use by response personnel during an emergency. Some chemicals found in laboratories are harmless, while others can be flammable, oxidizers, corrosive, toxic, or explosive. Firefighters may be confused by chemical names and not know by the name if a material is dangerous or not. Warning markings on smaller containers are not always conspicuous. While positive identification is most desired, determining a hazard class, or chemical family can identify the hazard generally associated with the chemical. The chart on "Hints for Hazardous Materials" can be helpful in determining relative danger of chemicals as an initial starting point. The DOT ERG can also be useful to first responders. Listed below are some common laboratory chemicals which become increasingly dangerous as they age.

Ethyl ether and other ethers are organic compounds that form explosive peroxides on contact with air. They are found in college, high school, research, and industrial laboratories. When a container of ether is opened, oxygen from the air forms a bond with the single oxygen present in each ether molecule and forms an organic peroxide. These peroxides are very unstable and become shock-, heat-, and friction-sensitive. Just moving or shaking a container can cause an explosion to occur which could seriously injure laboratory personnel, students, researchers, or emergency responders.

Ethers are also very flammable with wide flammable ranges (Figure 4.139). Fire is likely to follow an explosion of an ether container. Ethers were once used extensively as an anesthetic in hospitals, and while not highly toxic, could injure or impair responders. Ethers in laboratories should be dated when opened and discarded after 6 months in storage. Ethers stored longer than 6 months run the risk of peroxide formation.

Potassium metal, a metallic element, from family one on the periodic table of elements, is soft and silvery in color, and frequently found in high school and college laboratories. When shipped in transportation, it is found in a metal container stored under kerosene or naphtha to keep it from coming into contact with the air. While not air-reactive, potassium and other metals of family one can react to the moisture in the air. When encountered in labs, potassium and other metals are often stored in improper containers, such as mayonnaise or canning jars. This type of storage increases the dangers during an emergency should the glass containers be broken and expose the metal to the air and spill the flammable liquid also in the container. Like other members of family one, it is a dangerous fire risk and reacts violently with water to liberate hydrogen gas (which is highly flammable) (Figure 4.140). The heat from the reaction of the water and the potassium can be enough to ignite hydrogen.

When exposed to moist air it can also spontaneously ignite. Potassium is closely related to sodium and lithium metals, which are in the same family on the periodic table of elements. However, the similarity ends with their water- and air-reactive characteristics. Potassium metal becomes very dangerous in storage as it ages. Like ether, it also forms

Figure 4.139 Ethers are also very flammable, with wide flammable ranges.

Figure 4.140 Like other members of family one, it is a dangerous fire risk and reacts violently with water to liberate hydrogen gas (which is highly flammable).

explosive peroxides in long-term storage. Potassium can form peroxides and superoxides at room temperature and may explode violently when handled. Simply cutting a piece of potassium metal with a knife to conduct an experiment could cause an explosion. Its dangers far outweigh its usefulness in laboratory experiments in schools and should be replaced with sodium or lithium metals which also react with water but do not form explosive peroxides. Their degree of reactivity is less than that of potassium.

Picric acid is a type of chemical that is shipped and stored with a minimum of 10%–20% water in the container (Figure 4.141). While it is highly explosive when dry, it is classified as a 4.1 Flammable Solid, Wetted Explosive because of the moisture content in the container. It cannot be shipped when dry. As long as the moisture remains in the container, the compound is stable. Picric acid is a yellow crystal, which becomes highly explosive when it dries out, and is shocked or heated. The structural and molecular formulas of picric acid, (the common name for tri-nitro phenol), and its close relative, tri-nitro toluene (TNT), are very similar.

When dry, picric acid also closely resembles the explosive power of TNT pound for pound. Picric acid is another chemical which can be found in high school, college, and research laboratories. As it ages, the moisture which keeps it stable evaporates, and it becomes an unstable and highly explosive.

Benzoyl peroxide is a white, granular, crystalline solid, which is highly flammable, explosive, and toxic by inhalation. It is also a material which may explode spontaneously when it becomes dry.

Figure 4.141 Picric acid is a type of chemical that is shipped and stored with a minimum of 10%–20% water in the container.

Phosphorus is a wax-like crystal, transparent solid material. White or yellow (same material but can be called by either color), phosphorus is the most common, reactive, and dangerous. Red and black phosphorus can also be found in laboratories but do not possess the same dangers as white or yellow. White phosphorus is an air-reactive material that must be stored underwater or other liquids to keep it from spontaneously igniting. Like potassium and other metals, it is shipped in metal containers. However, many times it can be found in laboratories, especially high schools, in glass containers which can prove a significant hazard during an emergency. In addition to being air-reactive, it is also quite toxic at 0.1 mg/m^3 in air and is commonly used as a rat poison. It doesn't appreciably deteriorate with age but is very dangerous if not properly stored and handled and can cause serious fire and burns on people exposed.

Firefighting Operations

Firefighting operations usually involve the application of water through a hose line. Small fires in laboratories can be extinguished safely by using portable dry chemical fire extinguishers. When metals such as potassium are involved, Class D dry powder fire extinguishers should be used. Any other type will not be effective. Inserting a hose line into a laboratory chemical storage area can cause glass containers to break and mix chemicals together. Even a chemist couldn't tell you what the potential outcome would be if chemicals were mixed that are not normally placed together. Firefighters and other rescue personnel may encounter highly toxic and carcinogenic chemicals and mixtures along with flammable, water and air-reactive, and explosive materials. Great care should be taken when fires occur in laboratories. Run-off from firefighting can be very toxic and cause environmental damage, or at the very least danger to personnel and contamination of PPE. Some fires may be better left to the sprinkler system to extinguish or allowed to burn themselves out. To ensure the safety of personnel, some important steps should be taken in handling chemical fires and emergencies.

- Firefighters and other rescue personnel should never approach a chemical emergency scene or fire without SCBA and proper protective clothing. Care should be taken not to contact any chemicals or runoff with turnouts.
- During overhaul, extreme care should be taken to ensure firefighters are not exposed to chemicals. Also, runoff water, which may have become contaminated during firefighting operations, should be controlled.

Several states, including Minnesota, Nebraska Virginia, and Iowa, have developed plans to regulate chemical storage in public schools. The basic concept behind these programs is to make a one-time sweep through the states to remove and properly dispose of unwanted, dangerous, and aging chemicals. This requires the cooperation of many state and local agencies. For instance, the State Department of Education might be involved in the notification process by mailing letters to school districts introducing the program and asking science teachers if they have any chemicals that need to be removed. Inventories could then be compiled and forwarded to the State Fire Marshal's Office to coordinate the removal of the chemicals. Inventory lists may be divided into two categories, those that can be destroyed and those that must go to a hazardous waste landfill. State Fire Marshal personnel might work with State Police explosives technicians or local bomb squad experts to remove and dispose of explosive, flammable, and reactive chemicals. This could be accomplished by detonation using plastic or other commercial explosives to vaporize unwanted chemicals. Remaining toxic and carcinogenic chemicals could be collected at central locations throughout a state or city and packaged for shipment to a hazardous waste landfill. By developing a coordinated disposal program, the costs can be reduced as much as 50% or more depending on the location in the country.

In addition to removing dangerous chemicals from laboratory settings, it is important to educate teachers, laboratory workers, and research personnel on proper storage, use, and purchasing practices through training classes and literature. High school and college instructors should also be given assistance to develop experiments using safer chemicals that produce similar results but reduce the danger to students, teachers, and emergency responders. Through the cooperative efforts of school administrations, industry leaders, research organizations, and emergency response agencies, schools and laboratory settings can be made safer for occupants and response personnel. With proper training and preparation, emergencies involving hazardous chemicals in schools and laboratories can be safely and successfully handled (*Firehouse Magazine; Hazardous Materials Chemistry for Emergency Responders*).

Chemical Warfare Agents

The Sarin nerve agent attack that occurred in a Tokyo subway in 1994, and an earlier such terrorist incident in Matsumoto, Japan, added a new dimension to the threat of chemical agents (Figure 4.142). In the hands of terrorists, these agents pose a real threat to the population, and almost any public area is vulnerable. Emergency responders may be faced with hundreds and even thousands of casualties. Firefighters and EMS personnel may be unprepared to deal with chemical agents – and their victims – released in a terrorist attack. During the Tokyo attack, medical personnel and emergency responders didn't realize the source of the victims' medical problems until some time had passed; no decontamination was performed.

Chemical warfare agents were introduced during World War I and were used as recently as the Persian Gulf War. These agents had been stockpiled at seven sites in the mainland United States and Johnston

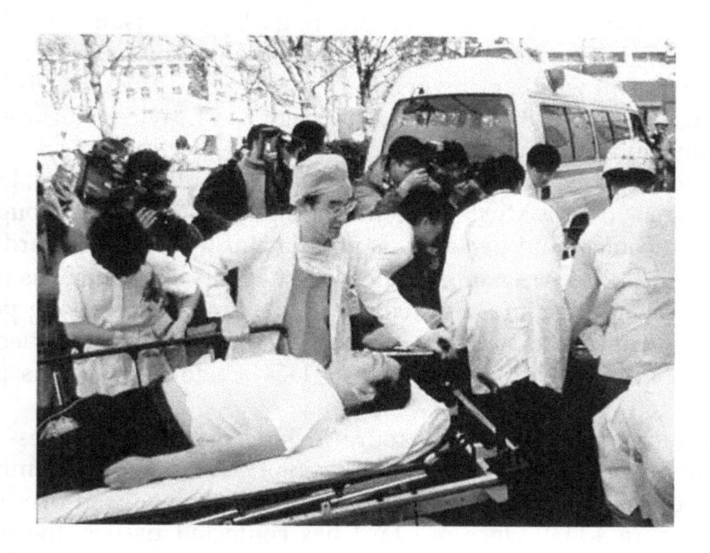

Figure 4.142 The Sarin nerve agent attack that occurred in a Tokyo subway in 1994 and an earlier such terrorist incident in Matsumoto, Japan added a new dimension to the threat of chemical agents.

Island in the South Pacific since World War I. The U.S. stockpiles of "unitary" chemical warfare agents have been destroyed as a result of an international treaty before 2004. (Unitary agents are those that act on their own without having to be mixed with another chemical to be activated.) Many of the aging munitions experienced leaks of chemical agents at some of the facilities, although to date no off-post releases of any significance have occurred. The U.S. Army and Federal Emergency Management Agency (FEMA) coordinates efforts to ensure public safety in the event an accident occurring involving any stockpiled chemical agents. This cooperative effort is the Chemical Stockpile Emergency Preparedness Program (CSEPP). Emergency responders in areas likely to be affected by a release are trained in chemical agent awareness, emergency medical treatment, and response procedures including PPE and decontamination. Chemical agents also have been found around the country in unexploded munitions buried on and off present and past military installations.

Blister Agents

There are two primary types of chemical agents: mustard and nerve. Mustard agents are also referred to as vesicants or blister agents because they form blisters when they contact the skin. The two types of mustard agents are sulfur mustard and nitrogen mustard. Nitrogen mustard is similar to sulfur mustard as far as its health effects but has slightly more systemic effects (affecting the entire body.) Mustard is often incorrectly referred to as "mustard gas." Mustard agent is a viscous oily liquid, light yellow-to-brown in color, with an onion, garlic, or mustard smell and it freezes at 57°F. Mustard is usually nonvolatile but may produce a vapor hazard in warm weather or when involved in a fire. The vapor is heavier than air, with a density of 5.4. Mustard agent is heavier than water and nonsoluble in water.

There are three basic types of blister agents: sulfur mustard, lewisite, and phosgene oxime. Mustard agent is designed to function through skin and tissue contact and generally is not a major inhalation hazard under normal conditions. Vesicants are persistent agents, which means they do not readily vaporize and will remain as contaminants for long periods. Mustard agents do not cause pain on contact and do not act immediately; instead, there is a dormant period of 1–24 h before symptoms present themselves.

Symptoms of mustard exposure include erythema (redness of the skin, sunburn type injury), blisters, conjunctivitis (eye inflammation), and upper respiratory distress; the symptoms may worsen over several hours (Figure 4.143). Once mustard has contacted tissues, the damage has already started, although there may not be any indication that contact has occurred. Mustard is highly soluble in fat, which results in rapid skin penetration. The lethal dose of liquid mustard applied to the skin

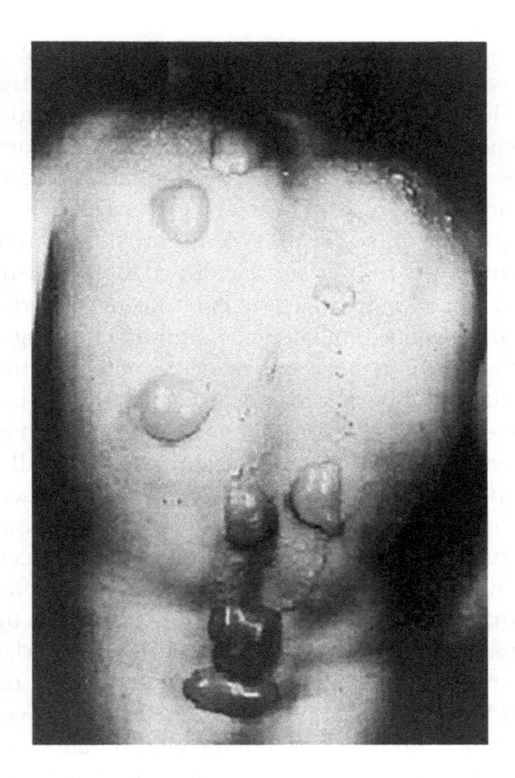

Figure 4.143 Symptoms of mustard exposure include erythema (redness of the skin, sunburn type injury), blisters, conjunctivitis (eye inflammation), and upper respiratory distress; the symptoms may worsen over several hours.

is about 7 g spread over 25% of the body surface area. The threshold for skin erythema and blistering is 10 μg/cm^2 (a microgram is one-millionth of a gram) deposited on the skin. The median lethal dose (LD_{50}, or lethal dose for 50% of the population) for ingestion of mustard is estimated as 0.7 mg/kg of body weight. Damage occurs primarily to the skin, eyes, and respiratory tract.

Vapor threshold doses of mustard that cause effects to the eyes are 0.1 mg/m^3 over 10–30 min of continuous exposure. Doses of 200 mg-min/m^3 may cause corneal edema (swelling and fluid build-up within the cornea), and blepharospasm (uncontrolled winking), leading to temporary blindness. Irritation of the nasal mucous membranes and hoarseness first occur at doses ranging from 12 to 70 mg-min/m^3. Lower respiratory effects, such as tracheobronchitis (bronchitis or inflammation of the trachea), tachypnea (excessively rapid respirations), cough and bronchopneumonia (inflammation of the lungs), begin to occur with doses exceeding 200 mg-min/m^3.

Mild skin erythema may be seen with doses of 50 mg-min/m^3. Severe erythema, followed by blistering, may begin at concentration-time profiles exceeding 300 mg-min/m^3. Warm, humid environments may cause earlier development of erythema and blisters at lower doses. The maximum safe doses for mustard have been established as 5 mg-min/m^3 for skin exposure and 2 mg-min/m^3 for eye exposure. The threshold limit value-time weighted average (TLV-TWA) for mustard is 0.003 mg/m^3. The immediately Dangerous to Life and Health (IDLH) measure for mustard is recommended at 1.67 mg/m^3. There is no known antidote for mustard agent. Mustard has been established as a human carcinogen.

Unless decontamination occurs within seconds after exposure to mustard, the results will be minimal. Hypochlorite solutions or large volumes of water are used to attempt to flush the agent from affected tissues. If decontamination is begun after symptoms start, it will have no effect. Mustard enters the body through the cells of the skin or mucous membranes and produces biochemical damage within seconds; no known procedure can reverse the process. Treatment of mustard exposure is much the same as for thermal burns and involves managing the symptoms and lesions (blisters). Mustard is rapidly transformed when it contacts the tissues in the body but is not found to be present in the blood, blister fluid, tissue, or urine. Mustard exposures are rarely fatal and patients recover over a period of months. In World War I, one-third of all U.S. casualties were the result of chemical exposure (sulfur mustard) but the fatality rate was only 5%.

Lewisite is a vesicant, believed to have been used by Japan against China in the 1930s. It has no other known battlefield uses. Phosgene oxime is also a vesicant, which has not been used on the battlefield. Mustard is found in artillery shells, mortar shells, land mines, and ton containers, which are similar to the ton containers used to ship and store chlorine. It is not likely that mustard would be used in terrorist activity because of the time it takes for mustard to act and its low volatility; however, if it were to be used, many hours would pass before the symptoms began to surface. This time delay would make diagnosis and determination of the incident source difficult.

Nerve Agents

Nerve agents are the most toxic of all chemical agents used in weapons and by terrorists. Nerve agents are nothing more than close relatives of organophosphate pesticides; however, military nerve agents are much more toxic. In fact, many of the nerve agents were discovered by chemists trying to make new pesticides. Pesticides generally work by interfering with the functions of the central nervous system, and nerve agents act by the same means. Some common organophosphate pesticides include malathion and parathion. Another group of pesticides called carbamates cause the same type of central nervous system damage as the nerve agents although they are not related chemically. A common carbamate is the pesticide Sevin. Nerve agents were discovered in the 1800s but their toxicity was not realized until the early 1900s.

Nerve agents under moderate temperature conditions are liquids. They are clear, colorless, and tasteless, and most are odorless. The primary military nerve agents of importance are GA (Tabun), GB (Sarin used in the terrorist attacks in Japan), GD (Soman), GF, and VX. GB agent is the most volatile even though the evaporation rate is less than that of water. GD has a greater evaporation rate then GA, and GA has a greater rate than GF. It is unlikely that GD should be used as a terrorist weapon because of the complexity of its manufacture. VX has a viscosity similar to light motor oil, and although it produces a slight vapor, it generally is not considered to be a vapor hazard unless the ambient temperature is very warm.

In addition to being inhalation hazards, all nerve agents are also absorbed through the skin and will travel through ordinary clothing. The LD_{50} of VX is much smaller than for GB, GD, GA, and GF. However, this is because the "G-Agents" may evaporate from the skin before they can penetrate. Tabun was first made in the 1930s by a German chemist. Sarin was also discovered in Germany about 2 years after Tabun. The German government manufactured large amounts of these nerve agents and stockpiled them during World War II but never used them. The Allies were not aware of the existence of the chemical nerve agents and had no protection or antidotes against them. After World War II, Soman, GF, and VX were discovered, manufactured, and stockpiled by the United States and the Soviet Union. The only known battlefield use of nerve agents was during the Iraq–Iran War when Iraq used nerve agents against Iran. The major concern presently with nerve agents is the manufacture and use by terrorist groups.

Unlike the effects of slow-acting vesicants, vapor exposure to nerve agents will cause symptoms within seconds to several minutes after contact. Large amounts of a nerve agent in vapor form will cause loss of consciousness and convulsions within seconds after one or two breaths. Nerve agent exposures on the skin will not present symptoms for varying periods of time depending on the amount of the exposure. Effects occur from several minutes to as much as 18 h. It takes time for the agent to penetrate the skin and reach the target organ(s). Nerve agents act by inhibiting the enzyme cholinesterase. The function of cholinesterase is to destroy the neurotransmitter acetylcholine.

Neurotransmitters are chemical substances released by a nerve impulse at the nerve ending. When released, they travel to the organ that the nerve stimulates. Once it arrives at the organ, the neurotransmitter combines with the receptor site the organ to affect the organ. For example, to move a muscle anywhere in your body, an electrical impulse originates in the brain and travels down appropriate nerves to the nerve ending near that muscle. The electrical impulse does not go to the muscle but causes the release of a neurotransmitter, which then travels across the very tiny gap between the nerve ending and the muscle to stimulate the muscle. The muscle reacts to this stimulation by moving. The neurotransmitter is then destroyed to prevent the stimulation of the muscle again.

If additional muscle movement is required, another nerve impulse causes the release of more neurotransmitter. The neurotransmitter acetylcholine is released by nerve endings and stimulates the intended organ. It is then destroyed by the enzyme acetylcholinesterase. As long as the acetylcholinesterase is intact, the body functions normally. If the cholinesterase is inhibited, the acetylcholine builds up and overstimulates the muscles, glands, and other nerves, which produce the symptoms exhibited by nerve agent exposures.

The cholinergic nervous system stimulates skeletal muscles (those that voluntarily move such as the arms, legs, trunk, an face.) Additionally the exocrine glands (lachrymal glands, nasal glands, salivary glands, sweat glands, and the glands that line the airways and gastrointestinal tract) are stimulated by acetylcholine. Smooth muscles (of primary importance are the muscles that surround the airways and the gastrointestinal tract) are also stimulated by the neurotransmitters. When the acetylcholinesterase is inhibited, the excess acetylcholine overstimulates all of these structures to cause involuntary movement in the skeletal muscles. Excess secretions develop from the lachrymal, nasal, salivary, and sweat glands and continue into the airways and gastrointestinal tract with resulting constriction of muscles in the airways that cause bronchoconstriction, similar to asthma. Constriction in the gastrointestinal tract causes cramps, vomiting, and diarrhea.

Using atropine as an antidote can reverse the effects of nerve agents. Atropine blocks the effects of the excess acetylcholine and is most effective for smooth muscles and glands but does not help the skeletal muscles. A drug called pralidoxime chloride (2-PAMCl) is used in conjunction with atropine to treat the skeletal muscles. Convulsions may also occur with exposures to nerve agents and diazepam (valium) is administered in some instances to help control the convulsions. Local protocols should be established for the administration of antidotes based on advice from the local EMS medical director (*Counter-Terrorism for Emergency Responders*).

Molten Materials

Sulfur (Figure 4.144)

Safety at a glance

- May contain inhalation risk from hydrogen sulfide gas.
- Loaded at 285°±15°, and solidifies at 250° over time in transit.
- Properties of molten sulfur change when it solidifies.

Figure 4.144 Sulfur may contain inhalation risk from hydrogen sulfide gas.

- Be aware of heat hazard in the liquid phase.
- In the event of a derailment and breach, leaking contents are at an elevated temperature.

Asphalt

Tank trucks hauling asphalt will be placarded with a "HOT" placard (Figure 4.145). Asphalt is a thermal hazard and an inhalation hazard. Firefighter turnouts with SCBA would be required for fighting fires and doing reconnaissance but personnel should avoid any contact with the liquid.

Firefighters should wear full protective clothing and positive-pressure SCBA with a full facepiece, as appropriate. Avoid using straight water streams. Water spray and foam (AFFF/ATC) must be applied carefully to avoid frothing and from as far a distance as possible. Avoid excessive

Figure 4.145 Tank trucks hauling asphalt are placarded with a "HOT" placard.

water spray application. Keep run-off water out of sewers and water sources (Hazardous Materials Chemistry for Emergency Responders).

CASE STUDY

PERTH AMBOY, NJ JUNE 24, 1949
ASPHALT PLANT EXPLOSION

Three men died with two buried alive under flaming asphalt as a crackling series of explosions destroyed a $500,000 asphalt plant here yesterday. The shriveled, tar-covered bodies of two volunteer firemen could not be recovered for several hours after they were blown into a ditch filled with boiling asphalt. A third victim, a workman, died of burns later. Eight others were injured, two critically. Black, greasy smoke rose hundreds of feet into the air over the ruined California Refinery Co. plant. It was visible as far away as Manhattan, 25 miles to the north. The first explosion let go at 2 p.m., and fire spread rapidly to adjoining stills and storage tanks. Then a 10,000-gallon asphalt tank blew 50 ft into the air, spewing its blazing contents (Lowell Sun Massachusetts 1949-06-24).

Biological Materials

Anthrax

Terrorism has been a hot topic throughout the emergency response community for the past several years. Acts of terrorism occurring within the United States are still a very real possibility. Bombings have been the weapons of choice in the past however, biological weapons present a credible threat for the future. In particular, anthrax is a likely choice of terrorists for a number of reasons. First of all, it is highly lethal with a mortality rate from 80% to 90% from inhalation exposure. Deadly concentrations are much less than for chemical nerve agents. Anthrax is relatively easy to produce and occurs naturally in many parts of the world, including the United States. It is also easy to deliver in its most deadly form, aerosol. Anthrax weapons can be produced at a fraction of the cost of other weapons of mass destruction, with equal effectiveness. One kilogram (2.2 lb) of anthrax can be produced for less than $50.00 and if properly aerosolized, under appropriate weather conditions, could kill hundreds of thousands of people in a metropolitan area.

Following the attacks on 9/11, terrorism in the United States took on a new twist, the "anthrax hoax." A rash of threats boasting the release of anthrax or the exposure of individuals has occurred in at least 17 states and continues to spread. The FBI reports that it investigates new anthrax hoaxes on a daily basis. Targets have included abortion and family planning clinics, federal and state buildings, churches, courthouses, public schools, dance clubs, and office buildings. Threat mechanisms have included mailed letters with a white powder inside boasting "you've just been exposed to anthrax," threat letters, notes written on walls in buildings, telephoned threats, and materials left in buildings. These incidents are much like bomb threats which have occurred over the years. The potential for anthrax is very real, but all of the events that have occurred so far have turned out to be a hoax. Each time an anthrax threat occurs, it must be taken seriously.

Response to these types of incidents causes much the same response from fire, police, and EMS as a real anthrax release would cause. Response resources are tied up for hours, building activities are disrupted, as well as traffic around the buildings and neighborhoods. Thousands of dollars are lost from business disruption and the cost of emergency response to the incident. Media coverage plays into the attention the responsible person(s) is trying to obtain and encourages copycats to create similar hoaxes in other locations. With the type of response the emergency community is giving to these hoax events, a terrorist wouldn't need to have anthrax, or for that matter, any other biological material to create an act of terrorism, just the threat serves the purpose.

On October 30, 1998, abortion clinics in Indiana, Kentucky, and Tennessee all received letters claiming to contain anthrax. The letters carried postmarks from Cincinnati. In Indiana 31 people, thought to have been exposed, including a mail carrier, were decontaminated with chlorine bleach. They were then transported to hospitals, decontaminated again and given antibiotics. Hospital workers who treated the "victims" were also given antibiotics. In Los Angles, 91 people were quarantined for over 8 h after a threat reported anthrax had been released into the air handling system of a federal building. Once again antibiotics were administered to those thought to be exposed.

A state office building in Wichita, KS was evacuated after an envelope was found in a stairwell with a white powder and note claiming it was anthrax. Between 20 and 25 people were believed to have been contaminated. Fifteen were decontaminated as a precaution. On December 14, 1998, a school district secretary opened a letter which read "You've been exposed to anthrax." Several days later, a U.S. Bankruptcy Court in Woodland Hills, California was targeted and over 90 people were administered antibiotics. Three days later, telephoned threats caused two Van Nuys, CA courthouses to be evacuated, resulting in 1,500 people being quarantined for several hours. It has been estimated the incidents in California alone have cost the taxpayers $500,000 per occurrence. The list of incidents and locations goes on and on, but the circumstances and outcomes are all very similar and they have all been hoaxes.

What is Anthrax? Anthrax, *Bacillus anthracis,* also known as "woolsorters" disease, is a bacterial illness which is common among farm animals but can also affect humans. Two thousand to five thousand cases of naturally occurring anthrax are reported every year throughout the world. It most commonly occurs in South and Central America, Southern and Eastern Europe, Asia, Africa, the Caribbean, and the Middle East. About five cases are reported in the United States each year and occur mostly in five states, Texas, Louisiana, Mississippi, Oklahoma, and South Dakota. Considered to be an "occupational disease," people most likely to contract anthrax naturally, work around animals, such as veterinarians, ranch and farmworkers, and people who work with animal carcasses, hair, and wool.

Anthrax spores are very persistent and remain viable for years in soil, dried or processed hides, skins, and wool. They can survive in milk for 10 years, dried on filter paper for 41 years, dried on silk threads for 71 years, and in pond water for 2 years. There is an island off the Scottish Coast that was used by the British for anthrax tests during World War II, that is so contaminated it still cannot be inhabited to this day. Anthrax spores are resistant to many disinfectants, drying, heat, and sunlight. Anthrax, as a disease has three forms, cutaneous, pulmonary or inhalation, and intestinal. It is in the spore form that anthrax is the most dangerous.

While most of the powders suspected of being anthrax during hoax events have been white in color, resembling talcum powder, anthrax spores resemble cinnamon or cocoa in color and consistency. Spores could, however, be mixed with another substance and present a different color. Therefore, identification cannot be based upon the physical appearance of a suspected substance. There isn't any particular smell associated with anthrax spores, so identification by odor will not be possible either. Some field identification tests are available to detect the presence of anthrax, but laboratory tests must be conducted to confirm the true identity of a suspected material.

Cutaneous anthrax is the most common naturally occurring form of the disease, and the most curable, if treated. Without treatment, it can be fatal in 5%–20% of the cases. Cutaneous anthrax is contracted by contact with a source of infection through openings in the skin, such as a cut or abrasion, or another opening. Exposure usually occurs on an exposed area of the face, neck, or arms. The odds of being infected by anthrax in this manner are approximately 1 in 100,000. The incubation period can range from 1 to 6 days. Within approximately 3 days of exposure, an elevated, reddened papule develops at the point of contact, which may be slightly itchy. Swelling may involve the entire limb where the pimple forms. Treatment centers around the administration of antibiotics; penicillin is preferred but, erythromycin, tetracycline's, ciprofloxacin, or chloramphenicol can also be used.

Intestinal anthrax occurs when undercooked meat from infected animals is eaten. This type of exposure rarely occurs in the United States. Symptoms from ingestion of infected meat begin with acute inflammation of the gastrointestinal tract. This is accompanied by nausea, loss of appetite, vomiting, and fever. These symptoms are followed by abdominal pain, vomiting of blood, and severe diarrhea. Approximately 25% to 60% of those exposed in this manner will die.

Inhalation anthrax (woolsorters' disease) is by far the most deadly and the most likely form a terrorist would use. This route of exposure involves the inhalation of dry anthrax spores. It is reported by the U.S. Army that the lethal dose of anthrax spores through inhalation would be approximately 8,000–10,000 spores. Spores lodge in the alveoli of the lungs, which provides a medium for the spores to germinate and grow. For the spores to enter the alveoli, they must be from 1 to 5 μm in size. Symptoms of inhalation anthrax resemble a common cold initially and progress to severe breathing problems and shock after several days. Fever, malaise, and fatigue are early symptoms which may be accompanied by nonproductive cough and chest discomfort.

These symptoms may be followed by a short period of improvement ranging from hours to 2–3 days. The improvement is then followed by abrupt development of severe respiratory distress accompanied by dyspnea (difficulty breathing), diaphoresis (profuse sweating), stridor (harsh vibrating sound which is heard during respiration when the of airway obstruction), and cyanosis (bluish or purplish skin discoloration as a result of oxygen deficiency in the blood). Shock and death will occur 1–2 days after the onset of symptoms. Inhalation anthrax is almost always fatal once symptoms have begun. Antibiotics have minimal effect on inhalation anthrax. Those who do recover from all types of the disease may develop an immunity. Second attacks from the disease are unlikely.

A vaccine is available for anthrax. Administration of the vaccine occurs in six doses, at 0, 2, and 4 weeks followed by boosters at 6, 12, and 18 months. After the initial six doses, an annual booster is required.

As a result of recent anthrax scares, the Weapons of Mass Destruction (WMD) Unit of the FBI has developed guidelines for emergency responders. These guidelines are designed to reduce the impact of a hoax event on the response system, potential victims, and the community.

- "Persons exposed to anthrax are not contagious and quarantine is thus not appropriate. All first responders should follow local protocols for hazardous materials incidents involving biological hazards. Upon receipt of a threat, a thorough hazard risk assessment should be conducted. Upon notification, the FBI will coordinate a risk assessment in conjunction with the health department and other authorities on biological agents to ensure timely dissemination of appropriate technical advice."
- "Any contaminated evidence gathered at the scene should be triple-bagged. Individuals should be advised to await laboratory test results, which will be available within 48h. These individuals do **NOT** need to be placed on chemoprophylaxis (antibiotics) while awaiting laboratory test results to determine whether an infectious agent was present."
- "The individual needs to be instructed that if they become ill before laboratory results are available, they should immediately contact their local health department and proceed immediately to a predetermined emergency department, where they should inform the attending staff of their potential exposure."
- "Responders can be protected from anthrax spores by donning splash protection, gloves, and a full-face respirator with high-efficiency particulate air filters (HEPA) (Level C) or SCBA (Level-B). Victims who may be in the immediate area and are potentially contaminated should be decontaminated with soap and water: no bleach solutions are required. A 1:10 dilution of household bleach (i.e., Clorox 5.25% hypochlorite) should only be used if there is confirmation of the agent and an inability to remove the materials through soap and water decontamination. Additionally, the use of bleach decontamination is **Only** recommended after soap and water decontamination and should be rinsed off after 10–15min. Technical assistance can be immediately provided by contacting the National Response Center (NRC) at (800) 424-8802."
- "If the envelope or package remains sealed (not opened), then first responders should not take any action other than notifying the FBI and packaging the evidence. Quarantine, evacuation, decontamination, and chemoprophylaxis efforts are **NOT** indicated if the envelope or package remains sealed. Also, anthrax will likely be visible as a powder or powder residue visible powder is a Section 2332a. it should be reported to the FBI immediately.

"This information is provided by the WMD Operations Unit of the Federal Bureau of Investigation and the National Domestic Preparedness Office (NDPO), in coordination with the Centers for Disease Control, the Department of Health and Human Services/Office of Emergency

Preparedness, and the U.S. Army Medical Research Institute of Infectious Diseases (USAMRIID). The NDPO was established to coordinate the Federal Government's efforts to prepare the nation's response community for threats involving weapons of mass destruction. Contact your local FBI office if confronted by a WMD threat."

When responding to a suspected terrorist threat or incident involving anthrax or any other WMD, the first concern needs to be containment. Remember the responsibilities of the first responder to any type of hazardous material, recognition/identification, isolation, notification, and protection. Limit the exposed area and the people allowed into the area. If the air handling system is involved, shut down the system. The entire building does not have to be evacuated unless it is suspected that the agent has been aerosolized and has gotten into the air handling unit, spreading it throughout the building. In the case of a letter or package without aerosolization, contain the incident to the room involved. Evacuation of the building will not be necessary.

Anthrax spores in a package or envelope do not present any particular hazard as long as no one touches, inhales, or ingests the material. Have the person who opened the package or envelope close it with the material inside. Have people potentially exposed keep their hands away from their face, and do not touch eyes, nose, or mouth. Hands should be washed with soap and water while waiting for decontamination facilities to be set up. Appropriate actions taken by first responders can reduce the impact these types of incidents will have on the facility, response personnel, and the community. Downplaying the incidents will help to reduce the number of copycat occurrences and the development of complacency among responders and the public.

Detection and Testing Outside of the military, there are few options for the detection of anthrax in the field. Current and developing test technology for anthrax can be classified into six general categories, antibody, "ELISA," blood, bacterial culture, microscope examination, and DNA.

- Antibody tests require a specimen form a powder, nasal swab, or surface swab. The sample is mixed with water and placed on a test strip, which contains anthrax-fighting antibody. A marking on the strip indicates positive. This test takes 15 min, is 95% accurate, and can be conducted in the field. The drawback is that swabs may not pick up bacteria.
- The ELISA test is similar to the test strip, but must be done in a laboratory. It takes several hours and is more accurate than the test strip.
- Blood tests require a sample of the infected person's blood, which is tested for anti-anthrax antibodies. The results are quick, but the

test must be repeated in a week or over several weeks to check for increases in antibodies. Increases would indicate exposure. Blood tests may miss early infection.

- Bacterial culture requires a sample of blood, powder, or swab, which is placed in a culture dish with nutrients specific to anthrax bacteria. If patches of bacteria appear, the test is positive. Test results take 2 days and are highly accurate. Anthrax spores can also be detected.
- Microscope examination involves samples of blood or swab being stained with a chemical. If the bacteria are boxcar shaped and turn purple, the test is positive. The test may not detect early infection and cannot detect dormant spores.
- DNA tests require a sample from blood, powder, or a swab, which is mixed with a chemical that is a mirror image of anthrax DNA. If the two interact, the sample is positive for anthrax. Testing takes 24 h or longer in an advanced laboratory setting.

Anthrax is a large *Bacillus* bacteria like many other germs that cause mild food poisoning and some which are harmless. Some quick field tests can identify if a *Bacillus* family member is present, but do not positively identify anthrax. A negative reading would be useless because there might not be enough bacteria in a sample to register. The use of nasal swabs that was widely reported on television is not a medically important test. It is merely an investigative tool to determine how many people might have been exposed and where. Spores might disappear by the time the swab test is conducted. Those prescribed antibiotics should not stop taking them just because a nasal swab is negative.

Several companies have developed hand-held assay field tests, but, according to the Centers for Disease Control (CDC), "the utility and validity of these assays are unknown." CDC has been asked to evaluate the sensitivity and specificity of the commercially available rapid, hand-held assays for *B. anthracis*. When the study is completed, results will be made available. Because of all the current activity, conclusions from this study are not expected in the near future. One company has a system, which it claims can detect anthrax in the field, but the number of spores present must be 10,000 or greater. At this level, the sample would be visible. In instances where there is no visible product, the test would be below the sensitivity of the product. Officials of the company say the CDC advisory was not directed at their product. The company claims the technology used in their product is also used by the military and FBI. So if the CDC finds their product is unreliable, then so would be the tests conducted by the military and FBI.

Research is underway on several fronts to develop technology for the field detection of anthrax. Some research is focusing on technology to allow for air monitoring of at-risk spaces such as subway systems. Those close to the research indicate it is still at least 2 years away from general use. These devices would be fully automated and behave similarly to smoke detectors. Another technology under investigation involves the use of DNA analysis in the field to identify anthrax. As recently as November 5, 2001 the Mayo Clinic announced they had a new DNA test that can detect the presence of anthrax in less than an hour. The device has not been approved by the Food and Drug Administration (FDA) because the number of anthrax cases has been too small for clinical trials to be conducted. Any tests conducted in the field should be followed up immediately by laboratory testing for confirmation. No actions should be taken concerning those exposed based solely on field testing.

Five-Step Process for Ruling Out Anthrax Spores Following the anthrax mail attack on the Eastern United States in September of 2001 emergency response organizations were inundated with "white powder" incidents across the country. All but a few of the calls, related to the actual anthrax attack on the East coast, were hoaxes. Responding to the extreme number of white powder incidents taxed the resources of many response organizations. There were so many requests for assistance that the FBI could not respond to all of the requests. They only wanted to be notified of "credible" incidents. Analytical laboratories that were asked to identify samples from incidents were overwhelmed by the number of requests. Field tests available at the time to test for anthrax spores were unreliable and fostered many false positives. Laboratory test equipment that could be used in the field to test for anthrax is very expensive and not available in most areas outside the military. Personnel at the Center for Domestic Preparedness (CDP) in Anniston, AL realized there was a need for a better way to deal with "white powder" incidents. CDP a component of the office of domestic preparedness is the Department of Homeland Security training center for WMDs.

They are the only emergency responder training center in the world that provides live agent training using sarin and VX nerve agents. The center opened on June 1, 1998, and has since trained 186,000 emergency responders through on and off-campus training opportunities. The facilities are situated on 62 acres at the site of the former Army Base Fort McClellan. Plans are in the works to acquire an additional 23.5 acres at the Fort with buildings that will be used for hands-on training. In addition to training, CDP also does WMD equipment evaluation using first responders to determine the usefulness of various items on the market. They obtain different types of equipment from different manufacturers and have emergency responders compare and rate them after using them under realistic conditions. Responders wear full PPE while using the equipment where appropriate. While I was there they were evaluating different commercial equipment for carrying nonambulatory patients out of incident hot zones, through decontamination and on to triage and treatment.

To deal with the problems associated with white powder incidents CDP assembled a working group in late December 2002 to attempt to develop a protocol for use by emergency responders at incident scenes. They wanted to pursue a science that would produce a low-cost detection method for anthrax spores in the field. Members of the working group included scientists, representatives from the fire service, law enforcement, EMS, hazardous materials response organizations, and emergency response educators. An analysis of responses to white powder/biological agent incidents yielded a number of weaknesses in preparedness at the local level.

- Lack of standardized protocols for response to biological incidents.
- Large numbers of responses exceeded local capabilities.
- Mid-sized and small communities lacked means of analysis to rapidly screen materials in the field to differentiate between potential hoax materials and real biological materials.
- Lack of safety guidelines for response to white powder incidents resulting in both overreaction and under reaction.
- Confusing and conflicting information from emergency responders resulted in skepticism by the public.
- Some detection equipment sold on the market is unregulated and invalidated.

Beginning the problem-solving process the Working Group established two objectives: (1) Determine how a typical American community's responders might quickly distinguish a hoax biological event from the real thing, given their current resources; and (2) develop a consensus field analysis protocol for responding to a potential anthrax letter. By the summer of 2003, as a result of the Working Group's efforts, a five-step biological field test system was proposed. First of all, it is important to note that the field test ***does not specifically identify anthrax spores or any other biological agent.*** What the field test system does is to "rule out" the presence of biological materials in a sample from a suspected white powder or biological agent.

Several factors regarding biological agents were addressed in the development of the testing system. Firstly, biological agents are living organisms. Biological materials typically survive and multiply within a pH range of 6.5–8.0 (Figure 4.146). Biological materials contain protein and other components such as carbohydrates, fats, and nucleic acids (DNA and RNA). Most biological materials, including anthrax spores, create a turbid (opaque or muddy) suspension when placed in water. Based upon the natural characteristics known about potential biological materials,

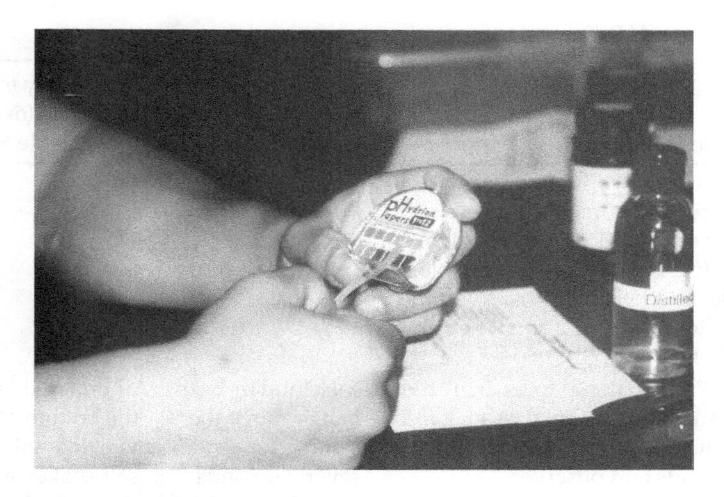

Figure 4.146 Biological materials typically survive and multiply within a pH range of 6.5–8.0.

the working group established test procedures designed to "rule out" the presence of biological agents. These procedures focus on particle size, water-solubility, pH, and protein content to assist emergency responders in evaluating unknown suspect powders (Table 4.1).

The Five Step Process is designed to be simple to conduct at an affordable cost, with a minimum of equipment. Supplies for the test kit include 3 mL glass or borosilicate vials, pH paper test strips with container comparison chart, protein test strips with container comparison chart, small disposable spatula or scoop, distilled water, a magnifying glass (10× or 20×), and a flashlight. Cost of the test process or kit was a side-focus of development efforts and was achieved in that the components of the test can be bought for less than $5.

A minimum of Level C personal protective clothing and respiratory protection with appropriate filters needs to be worn during sampling and testing procedures (Figure 4.147). Tests should be performed at the site where the powder or container of powder is located (within the hot zone) so that contamination does not spread. This testing process does not eliminate the need to take a sample or samples to be sent out for laboratory analysis. So, in addition to the test sample, response personnel should have necessary supplies for collecting a sample to go to the lab. Labs should have been contacted in advance of an incident to determine if they can test for biological materials and what they require in the way of sample collection, preservation, and sample size. Standing procedures that address packaging, chain of custody, and decontamination for suspect samples should be established with the local FBI or other law enforcement agencies in advance of an event.

Table 4.1 Five–Step Process Biological Field Test

Test to be Conducted	Possibly a Biological (or of Respirable Size)[a]	Not Likely to be a Biological (or not of Respirable Size)[a]
Step 1) Collect two samples: sample 1 for reference laboratory analysis[b]; a volume equal to 1 restaurant sugar packet (ca. 1 g) or more if available; sample 2 for field analysis; a volume similar to the size of a small pea or kernel of corn		
[a]Step 2) Place material to be analyzed in a dry ca. 3 mL clean glass vial and secure lid; shake vigorously for a few seconds and observe	Fine cloud or haze hangs above the sample for several seconds after shaking has stopped	All material falls to the bottom of the vial, like salt in a salt shaker, after shaking; air above the material is clear
Step 3) Remove lid, fill vial (ca.2 mL) with distilled water, and resecure lid; shake vigorously for 15 seconds and observe	Sample appears to mix with water but does not dissolve; liquid contents remain turbid or cloudy	Sample dissolves in water and becomes clear with or without larger particles settling to the bottom
Step 4) Remove the lid of vial and dip pH test strip into water; remove strip, wait 30 seconds, and read result on pH strip container	pH between 5 and 9	pH less than 5 or greater than 9
Step 5) Remove lid and dip 1 protein test strip; remove the strip, wait 30 seconds, and read result on protein strip container	Protein is present	Protein is not present

[a] The shaking of powder in the vial as described in Step 2 provides only an indication of particle size. This part of the test protocol does not provide an indication regarding the potential for the material being of biological origin.

[b] Standing procedures that address packaging, chain of custody and decontamination for suspect samples should be established with the local FBI in advance of an event.

Biological experts have validated the effectiveness of the field test procedures in the laboratory with a confidence level of 99%. That is to say, if all of the steps in the 5-Step Process produce negative results, there is a 99% chance that there is no anthrax or other biological agents. Live anthrax spores were used to test the effectiveness of the test kit in the laboratory setting.

CDP has developed a chart titled *Simple Biological Field Test System Confidence Factors* to assist response personnel in interpreting unclear

Figure 4.147 A minimum of Level C personal protective clothing and respiratory protection with appropriate filters needs to be worn during sampling and testing procedures.

test results and making decisions regarding powder incidents (Table 4.2). The shake test is used to determine particle size. Solubility, pH, and protein test results no matter in what order conducted can be evaluated by comparing results to Table 4.2. This will ascertain the approximate likelihood that the sample does not contain anthrax spores. Test scores that are in the red zone are considered to be dangerous and should be treated as such. Those in the amber and yellow zones should also be considered potentially dangerous. Green zone scores are considered very unlikely to be a biological threat and may be treated as such.

Once the Working Group reported its findings and the procedures were validated by biological experts, the next step was to determine how effectively emergency response personnel could utilize the procedure in the field. I was invited by CDP to witness the testing process and prepare this article for *Firehouse* to get the word out regarding this revolutionary new testing procedure. A group of hazardous materials technicians attending CDP the week of June 21–25, 2004 were asked to stay over an extra 2 days following their terrorism training class at the CDP to perform operational tests using the Five-Step Process.

Responders were provided with an overview of the test development process, instructions for conducting the tests, test kits, and PPE. Responders were divided into pairs and unmarked test samples were provided from the 15 most common substances used in white powder incidents. Harmless biological materials were also present among the test samples to show positive results from the testing process. For security purposes, those materials will not be identified here. Testing apparatus and procedures

Table 4.2 Simple Biological Field Test System Confidence Factors

	VERY HIGH THREAT- Credible threat, same as other confirmed attacks	HIGH THREAT- Credible threat, but no similarities to other confirmed attacks	Moderate Threat- Threatening, but assessed as possible copy cat	LOW THREAT- Same as other confirmed suspicious materials	VERY LOW THREAT- Other indications that there is no threat
HIGH RISK LOCATION (Federal or State government offices and courts, major businesses, location similar to other recent confirmed attacks, other hot spots such as transport hubs, women's health clinics or genetically modified organism research centers)	+Turbidity +Protein +pH	+Turbidity +Protein +pH	+Turbidity -Protein +pH Or -Turbidity +Protein +pH	+Turbidity -Protein +pH Or -Turbidity +Protein +pH	-Turbidity -Protein -pH Or +Turbidity -Protein -pH Or -Turbidity +Protein -pH
MODERATE RISK LOCATION (Local government offices and courts, controversial local businesses or institutions, schools)	+Turbidity +Protein +pH	+Turbidity +Protein +pH	+Turbidity -Protein +pH Or -Turbidity +Protein +pH	+Turbidity -Protein +pH Or -Turbidity +Protein +pH	-Turbidity -Protein -pH Or +Turbidity -Protein -pH Or -Turbidity +Protein -pH

(Continued)

Table 4.2 (Continued) Simple Biological Field Test System Confidence Factors

	VERY HIGH THREAT- Credible threat, same as other confirmed attacks	HIGH THREAT- Credible threat, but no similarities to other confirmed attacks	Moderate Threat- Threatening, but assessed as possible copy cat	LOW THREAT- Same as other confirmed suspicious materials	VERY LOW THREAT- Other indications that there is no threat
LOW RISK LOCATIONS (Private homes, small businesses)	+Turbidity +Protein +pH	+Turbidity +Protein +pH	+Turbidity -Protein +pH Or -Turbidity +Protein +pH	+Turbidity -Protein +pH Or -Turbidity +Protein +pH	-Turbidity -Protein -pH Or +Turbidity -Protein -pH Or -Turbidity +Protein pH

Source: Homeland First Response July/August 2003.

did not appear to cause any significant difficulty for response personnel. Each group of responders conducted tests on some of the 15 samples without protective equipment to become familiar with the procedures. Each responder was then given Level C protective clothing, cartridge respirator, two layers of gloves, and protective booties. An additional layer of latex gloves was placed on the outside of the butyl rubber gloves to make it easier to handle the test equipment and materials with gloved hands.

Once again there did not appear to be any major problems with response personnel conducting the tests using full Level C PPE. The complete test was conducted in a much shorter time than predicted by CDC staff. Results of the operational testing conducted by the emergency responders will be evaluated by scientists at the Aberdeen Proving Ground in Maryland. Initial findings from the laboratory and operational testing indicate the process is effective at ruling out anthrax specifically and other biological agents generally. Further tests are ongoing to ensure the best methods of employing the entire process are developed and documented prior to the release of the final reports (FEMA/CDP).

Ebola

Outbreaks of Ebola virus disease (EVD), (previously known as Ebola Hemorrhagic Fever or Ebola HF) have occurred over 20 times since the first documented outbreak in 1976. Ebola is named after the Ebola River in the African country of Zaire (now the Democratic Republic of Congo), which is where the first case of EVD occurred. Ebola is one of three viruses in the Filoviridae family of viruses. The other two viruses are Cueva virus and Marburg virus. Five subspecies of Ebola virus are known to exist; they are Zaire, Sudan, Tai Forest, and Bundibugyo. All four are believed to have caused disease in humans. The fifth subvirus, Reston, is believed to infect and cause disease in primates. Reston can also infect humans but has not been documented to have caused disease in humans at this time. Much investigation has been conducted to determine the host that carries Ebola virus, however, to date no natural host has been proven. Researchers, however, believe that the virus is carried by animals, with fruit bats being the most likely host. The World Health Organization (WHO) says infection of humans has been documented through the handling of infected chimps, gorillas, fruit bats, monkeys, forest antelope, and porcupines that have been found ill or dead in the rainforest. Pig farms in Africa are suspected to draw fruit bats and amplify outbreaks of EVD.

Once human infection has occurred, the Ebola virus can be transmitted from person to person through close contact with a person infected via contact with bodily fluids. Ebola virus spreads by direct contact (through broken skin or mucous membranes) with blood, secretions, organs, or other bodily fluids of infected people. It is also transmitted by direct contact with surfaces and materials (like bedding and clothing) and objects like needles and syringes contaminated with these fluids. Ebola virus is not spread through air or by water or by food. Ebola virus may be spread by handling wild animals hunted for food and contact with infected bats.

There is no evidence that mosquitoes or other insects can transmit Ebola virus. Only mammals, for example, humans, bats, monkeys, and apes have shown the ability to become infected with and spread Ebola virus. Healthcare providers caring for EVD patients and the family and friends in close contact with EVD are at the highest risk of getting sick because they may come in contact with infected blood or body fluids of sick patients. Health care workers in Africa have frequently been infected while treating patients with suspected or confirmed EVD. Infections occurred through close contact with patients when infection control precautions were not strictly practiced. In Africa burial ceremonies where mourners have direct contact with the body of the deceased person may also be a source of the transmission of Ebola virus. People infected with Ebola virus remain infectious as long as their blood and body fluids, including, but not limited to urine, saliva, sweat, feces, vomit, semen, and breast milk, contain the virus. This is usually thought to be 21 days, and those who have been exposed to an infected person or surfaces are watched for that amount of time, after which if they haven't developed symptoms, they are believed to be free of the virus. Men who have recovered from EVD can still transmit the virus through their semen for up to 7 weeks after their recovery.

The incubation period for the EVD is 2–21 days from the time of infection to when symptoms appear. Humans are not infectious until they develop symptoms. Initial symptoms are the sudden onset of fever (>101°F/38.3°C), fatigue, muscle pain, headache, and sore throat. This is followed by vomiting, diarrhea, rash, symptoms of impaired kidney and liver function and in some cases, both internal and external bleeding (e.g., oozing from the gums, blood in the stools). Laboratory findings may include low white blood cell and platelet counts and elevated liver enzymes. It may be difficult to distinguish EVD from other infectious diseases such as malaria, typhoid fever, and meningitis. Confirmation that symptoms are caused by EVD are made using the following investigations:

- Antibody-capture enzyme-linked immunosorbent assay (ELISA)
- Antigen-capture detection tests
- Reverse transcriptase-polymerase chain reaction (RT-PCR) assay
- Electron microscopy
- Virus isolation by culture

Samples from patients are an extreme biohazard risk; laboratory testing on noninactivated samples should be conducted under maximum biological containment conditions.

Supportive care-rehydration with oral or intravenous fluids and treatment of specific symptoms improves survival. There is as yet no proven treatment available for EVD. However, a range of potential treatments including blood products, immune therapies, and drug therapies are currently being evaluated. No licensed vaccines are available yet, but two potential vaccines are undergoing human safety testing.

In 2014, the African outbreak of EVD was brought to the United States for the first time. A U.S. citizen living in Dallas, TX was the first person diagnosed with EVD in the United States. The person had traveled to Liberia where the disease was contracted. Following unsuccessful treatment in a Dallas hospital, the victim died on October 8, 2014 of complications caused by EVD. Several health care workers who had been involved in the EVD patients treatment also contracted EVD but were successfully treated. Health care workers returning from treating EVD patients in Africa were also exposed and several became ill upon their return to the USA.

Other health care workers became sick while in Africa and were flown to the USA for treatment. One of the facilities that received Ebola patients was the University of Nebraska Medical Center (UNMC) in Omaha, Nebraska (Figure 4.148). UNMC has a ten-bed biocontainment unit which is designed and was commissioned in 2005 by the United States Centers for Disease Control (CDC) to provide first-line treatment for people affected by bioterrorism or extremely infectious naturally occurring diseases such as EVD. The facility in Omaha is the largest of its kind in the United States. Highly contagious and deadly infectious conditions that can be handled in the unit include SARS, smallpox, tularemia, plague, EVD, and other hemorrhagic fevers, monkeypox, vancomycin-resistant *Staphylococcus aureus* (VRSA), and multidrug-resistant tuberculosis. Safety measures are built into the biocontainment unit including air handling equipment, high-level filtration, and ultraviolet light, that prevent microorganisms from spreading beyond patient rooms. The entire bio unit is isolated from the rest of the hospital with its own ventilation system and secured access. A dunk tank is provided for laboratory specimens, and a pass-through autoclave are also in place to ensure that hazardous infections are contained. Also in the unit is a special sterilizer for laundry so that contaminated bed clothing is not removed from the unit. Patients are flown into Omaha's Eppley Airfield and transported to the UNMC Bio Unit, in an individual isolation unit also called a BIOPOD, by Omaha Fire Department EMS (Figure 4.149). Biocontainment unit staff receive specialized training and participate in drills throughout the year.

PPE for EVD patient care is utilized following guidelines developed by the WHO and the CDC. There are three key principles contained in the guidance from the CDC.

Figure 4.148 One of the facilities that received Ebola patients was the University of Nebraska Medical Center (UNMC) in Omaha, NE. (Courtesy: University of Nebraska Medical Center.)

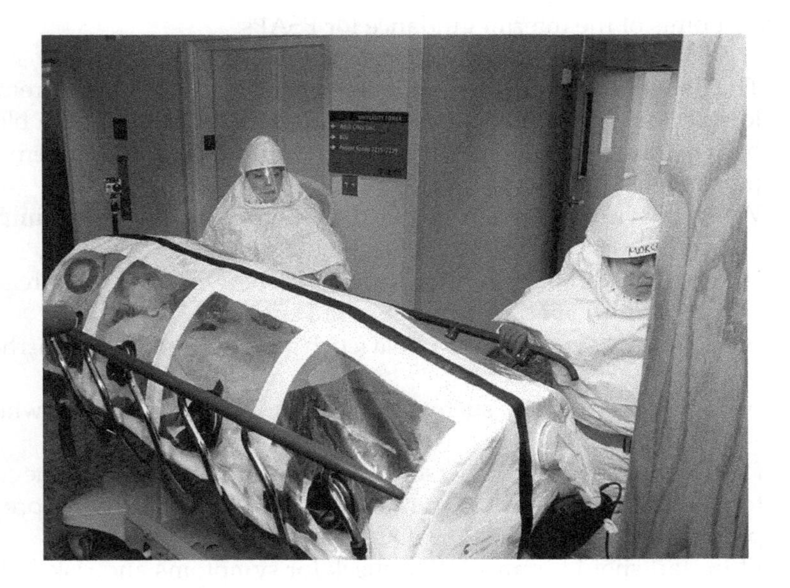

Figure 4.149 Patients flown into Omaha's Eppley Airfield and transported to the UNMC Bio Unit, in an individual isolation unit also called a BIOPOD, by Omaha Fire Department EMS. (Courtesy: University of Nebraska Medical Center.)

- Prior to working with Ebola patients, all health care workers involved in the care of Ebola patients must have received repeated training and have demonstrated competency in performing all Ebola-related infection control practices and procedures, and specifically in donning/doffing proper PPE.
- While working in PPE, health care workers caring for Ebola patients should have no skin exposed.
- The overall safe care of Ebola patients in a facility must be overseen by an onsite manager at all times, and each step of every PPE donning/doffing procedure must be supervised by a trained observer to ensure proper completion of established PPE protocols.

CDC has issued interim guidance for 911 Public Safety Answering Points (PSAPs) for Management of Patients with Known or Suspected EVD in the United States. This guidance is intended for Managers of PSAPs, EMS agencies, EMS systems, law enforcement agencies, and fire service agencies and to include individual emergency medical service providers (including emergency medical technicians (EMTs) paramedics, and medical first responders, such as law enforcement and fire service personnel).

Key points of the interim guidance for PSAPs:

- The likelihood of contracting Ebola in the United States is extremely low unless a person has direct unprotected contact with the blood or body fluids (like urine, saliva, feces, vomit, sweat, and semen) of a person who is sick with EVD.
- When the risk of Ebola is elevated in their community, it is important for PSAPs to question callers about:
  - Residence in, or travel to, a country where an Ebola outbreak is occurring (Liberia, Guinea, Sierra Leone);
  - Signs and symptoms of Ebola (such as fever, vomiting, diarrhea); and
  - Other risk factors, such as direct contact with someone who is sick with Ebola.
- PSAPs should tell EMS personnel this information before they get to the location so they can put on the correct PPE following proper procedures, as described in CDC's guidance.
- EMS staff should immediately check for symptoms and risk factors for Ebola. Staff should notify the receiving health care facility before EMS arrives with the patient.

Emergency medical service guidance for prehospital EVD response issued by CDC. Emergency medical personnel included in this guidance are prehospital EMS, law enforcement, and fire service first responders.

The single most important aspect of the prehospital response to EVD is ensuring that personnel wear the proper PPE and are fully trained in donning and doffing procedures. Advanced planning for EVD response and conducting drills are critical. PPE guidance for prehospital EVD response is the same as for hospital health care workers. Except for the PAPR, all PPE mentioned below is single-use, or in other words, disposable. PPE should ensure that no skin is exposed when the PPE is fully donned. If a PAPR is used a full faceshield, helmet or headpiece should also be used. Any reusable helmet or headpiece must also be covered with a single-use hood that extends to the shoulders and fully covers the neck and is compatible with the selected PAPR.

Recommended PPE for use by trained personnel includes (Figure 4.150): a positive pressure air-purifying respirator (PAPR) or N95 respirator in combination with a single-use surgical hood extending to the shoulders and single-use full face shield; fluid-resistant or impermeable gown that extends to at least mid-calf without integrated hood; nitrile examination gloves with extended cuffs; fluid-resistant or impermeable boot covers that extend to at least mid-calf; and a fluid-resistant or impermeable apron that covers the torso to the level of the mid-calf if EVD patients have vomiting or diarrhea.

Preparation for the care and transport of EVD patients by emergency responders requires advanced planning, training, and drills to ensure the best care of the patient and the safety of response personnel. 911 and dispatch personnel play a critical role in gathering patient information and transmitting that information to the responders. The information provided in this article is just an overview and by no means meant to

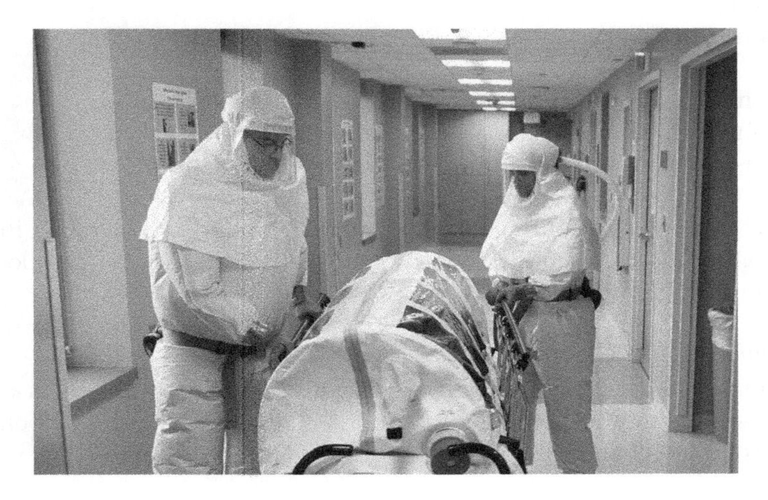

Figure 4.150 Recommended PPE for use by trained personnel. (Courtesy: University of Nebraska Medical Center.)

be a complete coverage of CDC Interim Guidance. Organizations need to obtain complete information from the CDC and other resources such as the World Health Organization, University of Nebraska Medical Center, and the National Institutes of Health to develop plans, standard operating procedures, and training to deal with Ebola and other infectious diseases. While Ebola is a very serious disease, handling and treating patients can be accomplished safely with proper preparation. Thanks to UNMC Senior Media Relations Coordinator Taylor Wilson for photographs, information sources, and News Release updates on UNMC's Biocontainment Unit and patients being treated there (*Firehouse Magazine*).

Confined Space

OSHA's 1910.134 is the impetus for using a multi-gas monitor to perform atmospheric testing prior to entering a confined space. When monitoring the atmosphere in a confined space, there are several important issues that need to be considered, reviewed, and managed. One of the major issues centers on air quality and what you are breathing, both prior to entry and during the occupation of a confined space. You need to know what the oxygen content of the atmosphere is and whether there are explosive or toxic gases that could threaten the safety of the environment or, perhaps, more importantly, your life. When properly used and maintained, gas detection monitors will protect both.

Not only do you need to monitor the atmosphere of your confined space to protect your life but also OSHA requires you to do so.

To understand exactly what a confined space is, let's look at OSHA's definition:

Confined space

1. has adequate size and configuration for employee entry,
2. has limited means for access or egress, and
3. is not designed for continuous employee occupancy.

A few examples of confined spaces could be underground vaults in the telecommunications industry, aeronautical fuel tanks, sewers, silos, or coal mines.

Permit-Required Confined Spaces

The term "permit-required confined space," as defined by OSHA, is a confined space that meets the definition of a confined space and has one or more of these characteristics:

1. contains or has the potential to contain a hazardous atmosphere,
2. contains a material that has the potential for engulfing the entrant,

3. has an internal configuration that might cause an entrant to be trapped or asphyxiated by inwardly converging walls or by a floor that slopes downward and tapers to a smaller cross-section, and/or
4. contains any other recognized serious safety or health hazards.

Once you have identified your work area as a confined space, you should refer to OSHA's recommendations for monitoring the air quality of your space.

Atmospheric Testing of Confined Spaces

OSHA standard 29 CFR 1910.146 (c) subsection (C) states: Before an employee enters the space, the internal atmosphere shall be tested, with a calibrated direct-reading instrument (Figure 4.151), for the following conditions in the order given:

1. oxygen content,
2. flammable gases and vapors, and
3. potential toxic air contaminants.

Figure 4.151 OSHA standard 29 CFR 1910.146 (c) subsection (C) states: Before an employee enters the space, the internal atmosphere shall be tested with a calibrated direct-reading instrument.

Additionally, subsection (D) states: There may be no hazardous atmosphere within the space whenever any employee is inside the space.

This standard is the impetus for using a multi-gas monitor to perform atmospheric testing prior to entering a confined space. It also clearly dictates that continuous monitoring of the space must take place for as long as the confined space is inhabited.

If hazards are found in the space through the utilization of a multi-gas monitor, OSHA standard 29 CFR 1910.146 (c) subsection (E) must be followed. This standard reads: Continuous forced air ventilation shall be used, as follows:

1. an employee may not enter the space until the forced air ventilation has eliminated any hazardous atmosphere;
2. the forced air ventilation shall be so directed as to ventilate the immediate areas where an employee will be present within the space and shall continue until all employees have left the space; and
3. the air supply for the forced air ventilation shall be from a clean source and may not increase the hazards of the space.

Choosing Your Confined Space Monitor

When choosing a monitor to test and continuously monitor a confined space, you should take into consideration several attributes of the monitor and be sure to accessorize accordingly.

First and foremost, you will need a multi-gas monitor capable of monitoring for all of the OSHA-required hazards: oxygen, flammable gases, and potential toxic air contaminants that may be present as a result of the processes that take place in or around the confined space. Next, you should consider a monitor that has either an internal or external pump that is capable of properly drawing the air sample back to your fresh air monitoring point during initial testing of the space. The monitor should also have the capability of continuously monitoring the occupied space to ensure the workers' continued safety. Other equipment such as sampling probes, durable carrying cases, and rechargeable batteries can be complementary accessories as well.

Most of today's monitors are equipped with bright visual and loud audible alarms to warn of potential hazards. An internal data logger will help you to comply with the documentation of your spaces' hazards. A data logger is a device containing a microprocessor that stores information electronically taken from an instrument. The levels of all hazards being monitored can be downloaded from the data logger to a computer or printed for reference and recordkeeping activities.

1. Event-logging mode, which records information when an incident or event occurs.

2. *Recharging.* Docking systems also can be used to charge monitors when not in use. This will ensure the monitor is fully charged the next time it is used.
3. *Instrument diagnostics.* Automated maintenance systems may include technology that provides a means for diagnosing potential problems with your monitor, such as low or marginal sensor life and date of the last calibration, along with the number of days until the next calibration is due.

Training Considerations

Some manufacturers of gas monitors also provide hands-on training to ensure that you use and maintain your instruments properly (Figure 4.152). These courses are designed to educate participants about the basics of proper gas monitor use and management. When looking for hands-on training, some of the more valuable topics you will want to learn about include:

1. *Hazardous gases.* Learning the common gases encountered in confined spaces, including information about deficient (asphyxiation hazard) or enriched (explosive hazard) oxygen levels, will prove valuable in your overall understanding of a confined space.
2. *Permit-required confined spaces.* Reviewing the regulations enacted under 29 CFR 1910.146 and learning how to utilize gas detection to meet the requirements will ensure compliance with OSHA's mandates.
3. *Sensor technology.* Learning about and understanding the various types of sensors used in confined spaces, such as catalytic diffusion,

Figure 4.152 Some manufacturers of gas monitors also provide hands-on training to ensure that you use and maintain your instruments properly.

electrochemical, and infrared sensors, will help you make an educated decision on instrument/sensor selection for your particular application.
4. *Instrument review.* Learning about the functions and applications of complete lines of portable instruments, including calibration stations and docking systems, will further help with purchasing decisions.
5. *Calibration and maintenance.* Learning to properly calibrate and care for your instruments and how to troubleshoot instrument problems and replace sensors and battery packs will give you the knowledge to be self-sufficient in maintaining your instrument inventory.
6. *Hands-on operation.* Learning to use a variety of portable instruments will give you the confidence to operate your instruments at peak efficiency to make certain you are protected at all times.

Proper Management of Ongoing Maintenance

Another very important aspect of managing your confined space gas monitor program is the proper management of ongoing maintenance. The best way to be certain your monitor is in peak shape is by utilizing the services of a manufacturer's Factory Service Center. Using factory-trained service technicians ensures the servicing of your monitor is performed by individuals qualified to work on it. Some of the key services to look for are:

- *In-house calibration and service.* This service will ensure your confined space monitor is calibrated and serviced by professionals.
- *Maintenance and warranty repair.* Performing routine maintenance and utilizing warranty repair services are critical to keeping your monitor fleet up and running.
- *On-site mobile service and repair.* Where available, this service will bring an authorized technician to your site to perform maintenance and repair service on the spot, eliminating downtime.
- *Instrument rentals and leasing.* Having the option to lease or rent an instrument proves very beneficial, especially in cases of shutdowns or planned maintenance where you may need more instruments than you own in-house to perform the work in a timely manner.

The combination of the right instrument, the proper training, and the services that complement your monitors will help make managing your confined space program easier.

Agriculture

Today's modern farming methods have brought new dangers that arise from farmers entering confined areas where oxygen levels may be inadequate or where toxic gases are present. When entering a confined area such as a manure pit, silo, grain bin, or an inadequately ventilated building, a farmer may be at risk of being overcome by gases or dust, which can cause permanent lung damage or death. Gases in manure pits and silos can quickly kill an unsuspecting farmer or an untrained rescue worker who enters the area without adequate protective equipment.

Farmers entering grain bins while the bin is being emptied may be taking an unnecessary risk of being crushed or suffocated by flowing grain. Working in grain bins without proper respiratory equipment to filter dust and molds increases a farmer's chances of developing a respiratory disease. Farmers working in dust-laden buildings run the risk of developing Farmer's Lung, a disease that permanently damages lung functions.

While most farmers are aware of the dangers of poisonous gases and flowing grain hazards, statistics from the past 4 years show that three to five Michigan farmers are killed each year in accidents involving these hazards. The occurrence of respiratory diseases among farmers cannot be accurately measured, but they are a concern among the rural population.

The intent of this publication is to make the reader aware of the risks associated with entering a confined space area and to provide information about risk reduction techniques for farm and orchard owners.

Farm Site Gases

There are several gases around farm sites that pose a risk to farmers, the three most common in confined space areas are hydrogen sulfide, ammonia, and carbon dioxide.

Hydrogen sulfide (H_2S) is formed during manure decomposition. It is toxic, combustible, and because it is heavier than air, it dissipates oxygen and can suffocate an unsuspecting farmer. Hydrogen sulfide also has a distinctive "rotten egg" stench that dulls the sense of smell, giving the farmer a false sense of security because the original odor disappears as exposure time increases. The gas irritates the eyes and respiratory tract. In low concentrations, hydrogen sulfide has been reported to cause headaches, nausea, and dizziness prior to the individual succumbing to the gas.

Ammonia (NH_3) is a suffocant and a toxic gas with a distinct, sharp penetrating odor. Prolonged exposure to ammonia or exposure to high concentrations of the gas can cause ulceration of the eyes and severe irritation to the respiratory system.

Carbon dioxide (CO_2) is a colorless, odorless suffocant that is produced during decomposition and respiration of plant materials. Excess carbon dioxide in a confined space depletes oxygen levels needed to sustain life. At low levels (CO_2 levels at 3%–6%), the individual may feel drowsy and develop a headache. Death from suffocation may result when carbon dioxide levels are 30% or greater (*Firehouse Magazine*).

Nuclear Magnetic Resonance (NMR)

Hazards of the "Invisible Force"

Powerful magnets are in use at research universities, medical centers, imaging centers, and industry. These magnets produce an "Invisible Force" that cannot be detected by the human senses but can be hazardous to firefighters and other emergency personnel if they do not understand the dangers that are present. In terms of the "Invisible Force" or magnetic field itself, there are no ill health effects from short-term exposure. However, this magnetic field produced by medical magnetic resonance imaging (MRI) and research nuclear magnetic resonance (NMR) devices can create an unexpected hazard to response personnel. Magnets attract metal objects. Because of the power of large magnets metal objects brought into an MRI or NMR room can be "ripped" from the hands or body of the person or persons with the metal objects and drawn into the magnet at 45 miles/h (Figure 4.153).

A small child receiving an MRI at a medical facility was killed when the magnet pulled an oxygen tank into the patient area of the unit hitting the child in the head. Fire extinguishers with metal tanks, firefighter tools, oxygen tanks, SCBA with metal tanks or metal parts of the regulator or harness, belt buckles, steel-toed shoes or boots, and anything else metal can be dangerous if taken into magnet rooms. The NMR magnets are more powerful than the electromagnets used to move old cars at the junkyard. Magnets can also damage certain types of credit cards, ID cards, watches, and calculators if taken into the magnet room. They may also affect pacemakers and other metal implants in the human body. Pay attention to warning signs placed on doors and in the area of powerful magnets. Read the directions on these signs and take them seriously.

These magnets are always turned on and cannot easily or quickly shut down during an emergency situation. Be careful when entering magnet rooms as there may be many obstacles such as tanks, piping, wires, and other apparatus associated with the magnet that can create trip and fall hazards as well as overhead dangers. When the magnet room is constructed great care is taken to make sure none of the construction materials contain ferrous (iron) metals that would be impacted by the magnets force. Sprinkler piping has to be copper or plastic. Fire extinguishers located in or near these magnet rooms need to have containers that are not ferrous metals. Specially designed extinguishers are commercially available.

Operation of one of these high-powered magnets requires the use of cryogenic liquid helium and liquid nitrogen. Both of these liquids are very cold and contact with the liquid can result in frostbite or solidification of body parts. There is no protective equipment that can be worn to protect the skin from the effects of cold liquids. The gases produced as the liquids

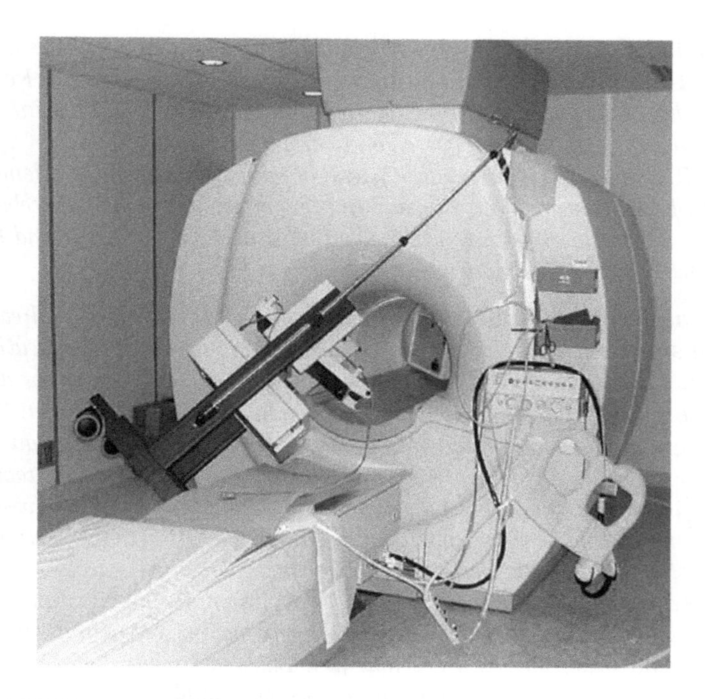

Figure 4.153 Because of the power of large magnets, metal objects brought into an MRI or NMR room can be "ripped" from the hands or body of the person or persons with the metal objects and drawn into the magnet at 45 miles/h.

warm can cause asphyxiation to those present by displacing the oxygen in the area in the event of a release. Wearing SCBA will protect the respiratory system and prevent asphyxiation. Magnet rooms are equipped with low oxygen warning monitors to warn of an oxygen deficiency. The only way to "shut down" the magnet is to activate the Quench button located in the magnet control room. This evacuates the nitrogen and helium from the magnet and in effect shuts it down. Quenching should only be done under the advice and consent of the magnet operator as it is an expensive operation to refill the magnet. Magnets provide an important function in medical treatment and research. Response personnel should be aware of the potential hazards of these powerful magnets and learn how to work safely around them during an emergency to avoid serious injury or death.

Water Injection for Liquid Propane (LPG) Leaks

Author's Note: *Ron Gore, retired Jacksonville, FL Fire Department, the "Godfather" of hazmat response in the American Fire Service, continues to impact the fire service today, 41 years after the Jacksonville team went* *operational. Ron Gore passed down the procedure for stopping leaks with liquid propane to Ron Huffman during a class at Safety Systems at Ron's Florida training facility. Ron Gore told Ron Huffman he did not invent the procedure, it had been passed on to him by someone else.*

Ron Huffman is a Facebook friend of mine (now a real-life friend!), and I became aware of his propane training efforts through his Facebook posts. This was something I had heard of in the past but never had the opportunity to learn more about.

***Hazmatology: The Science of Hazardous Materials** had already been submitted to the publisher and was in the process of publication when I heard about Ron Huffman's efforts. I contacted my editor and asked if it was too late to include this in Volume Four because I felt this was groundbreaking training that responders needed to know about; another tool for the Hazmat Tool Box. She told me I could, so I contacted Ron and he said he would put on a demonstration for me. Ron lives in Indiana, so I made arrangements for the Richmond, IN Fire Department to host and provide the demonstration for their personnel as well. Several years ago I had written a story about an explosion in Richmond in 1968 and had visited Richmond. This would give me an opportunity to see my friend Chief Jerry Purcell and Jack Bales, a survivor of the 1968 explosion, and make new friends at the Richmond Fire Department.*

Liquefied Petroleum Gases

Propane, as it is commonly called, is actually a mixture of several liquefied petroleum gases (LPGs). LPG includes liquid mixtures of volatile hydrocarbons propene, propane, butene, and butane. A typical commercial mixture may also contain ethane and ethylene, as well as a volatile mercaptan, an odorant added as a safety precaution as these materials are naturally odorless. This leak stopping procedure works on any of these mixtures as well as on pure compounds.

These gases are liquefied for commercial purposes. More of a gas can be transported and used in a liquefied form than as a compressed gas alone. LPG is stored and shipped at ambient temperatures. That is to say, the temperature around the container is the same as the temperature of the liquid inside. Pressure within the container keeps the gas as a liquid. If the liquid is released from the container, all of the liquid will very quickly be converted to a gas. One gallon of propane will expand to over 270 gallons of highly flammable gas. Generally, hydrocarbon compounds do not mix with water and are lighter than water, so liquid hydrocarbon compounds will float on top of water. It is the physical characteristics of water and LPG that allow for stopping liquid leaks in containers by injecting water into the containers.

Below is an overview of the process, as provided by Ron Huffman. This information is intended as an informational presentation and in no way provides training on the use of water injection or flaring systems or other associated procedures. It is up to your department to determine the need for this process and the associated equipment. Once a need is determined, personnel who will utilize water injection and flaring systems along with other procedures dealing with LPG must be thoroughly trained in utilizing these procedures and equipment by a knowledgeable instructor. Like any other tool in the Hazmat Tool Box, it will not be necessary or required in all instances. Personnel must know when use is proper and when use is not needed or too dangerous to personnel to use. Just because you have it does not mean all encounters with LPG will require its use. LPG is a flammable gas and appropriate precautions need to be taken, such as controlling ignition sources and monitoring lower explosive limit like any other flammable gas or vapor.

Emergency Water Injection – 101 Written By Ronald D. Huffman

When something happens and you're faced with an uncontrollable liquid propane leak you have very few options. Depending on the size of the container and the volume of the release, a liquid propane leak can quickly create a large hazard area requiring large area evacuations, limit your ability to safely extricate trapped accident victims, or identify the location of a leak.

Injecting water into a propane tank is a leak management option that most responders and suppliers have not considered. Water injection, just like all response tactics, requires special knowledge, identified tactical objectives, the proper tools, and training (Figure 4.154). Water injection has been used in some areas of the country for years, but for some reason, the process has not been adopted throughout the fire service and propane industry.

So now you have questions: What is it? How does it work? What type of incident would necessitate the use of Water Injection as a tactic? When should it not be used? What about the proper response equipment and training? ***First, what is it?*** Water injection is the process of forcing water into a tank or cylinder using a water supply capable of producing more pressure than the propane tank's static pressure. If you look at Table 1 you see that temperature dictates pressure. Under normal atmospheric temperatures, most fire apparatus are more than capable of producing the pressures needed.

How does water injection work? Propane exists in a pressure vessel in one of two states, liquid and/or gas. Liquid propane weighs a little more than 4 pounds per gallon and will remain in the lowest area of the tank regardless of the tanks orientation. This area is commonly known as the wetted space while the vapor space or tank surface area above the liquid is known as the dry or non-wetted space. If a leak occurs in the vapor space,

Figure 4.154 Instructor Ron Huffman teaching Richmond, IN firefighters and hazmat team members about the LPG water injection system.

vapor is released and conversely if a leak occurs in the wetted space, liquid propane is released.

Water weighs a little more than 8 pounds per gallon (approximately twice the weight of propane) and when injected into a propane tank will sink below the liquid propane lifting and replacing it at the bottom of the tank. Once the water rises above the leak opening, the propane leak will be converted to a water leak removing the expanding hazard. As long as water is continually injected equal to the volume of the leak, the tank will continually leak water and not propane. Sounds simple, right? It is. But,

Product Temperature	Minimum Pump Pressure
–10° F	50 psi
0° F	50 psi
10° F	50 psi
20° F	70 psi
30° F	70 psi
40° F	100 psi
50° F	100 psi
60° F	150 psi
70° F	150 psi
80° F	150 psi
90° F	170 psi
100° F	220 psi
110° F	250 psi

injecting water into a propane tank can provide response agencies with a capability that while useful can be dangerous if done incorrectly.

Incident #1, The tones go off and you're dispatched to a motor vehicle accident involving a bobtail. As you arrive you're faced with a passenger car wedged under the bobtail, and the front passengers are trapped and need extricated. The accident has damaged the bobtail causing a liquid leak that's producing a large vapor cloud exposing the trapped victims and responders to the very real possibility of a fire.

Incident #2, You're dispatched to the County Fair and a food vender has two 100-pound tanks and one is leaking through the bottom. The visible vapor cloud is extending approximately 50 feet downwind into the midway. We'll come back to these a little later.

What Are the Advantages of Water Injection?

Water injection provides the ability to convert a liquid propane leak into a water leak stopping vapor production and providing time to mitigate the incident. Other major benefits include, better visibility to find the source of a leak, provides a safer environment for rescue and may provide containment options that allow you to move the container to a safer area (Figure 4.155). Under the right circumstances water injection can also be used to force liquid propane up so that the tank can be evacuated through other valve options. This tactic must only be attempted by trained and experienced personnel and only on a tank that has lost its pressure due to auto refrigeration.

Is water injection safe? Yes, when the proper protocols are followed that include identifying the tanks liquid level and pressure prior to starting operations and continuously monitoring throughout the process.

Can it be done wrong and cause problems? Absolutely, ranging from rapidly increasing tank pressure, adding stress to an already compromised tank, hydrostatic rupture and crushing victims' due to the added weight to name a few.

Will water damage the propane truck or tank? No, water will not harm any component of a tank. MC331 vessels are regularly hydrostatically tested with water to 1 ½ times their normal working pressure. When a tank is tested the water is removed and methanol is normally injected to absorb any remaining water. Once a propane tank that has had water injected to mitigate an emergency it will require service by a qualified repair shop after the incident.

When not to inject water into a propane tank.

- When a tank is in a position that would not allow the propane to be floated above the leak location prior to over filling.
- Relief operation and no liquid leak.
- Tank volume more than 80% unless closely monitored.

Figure 4.155 Richmond firefighters and hazmat team members practice water injection on a propane tank leaking live LPG.

- To off load liquid when the vessel is on its side. The weight of a tank that is half full of water is as heavy as a full tank of propane.

Water Temperature and Special Hazards

Caution must be used when injecting water into a product that is subject to pressure changes due to temperature. Hose lines laying in the sun especially on a roadway can absorb a lot of heat and water circulating in the fire pump can be raised to boiling if not managed properly. The introduction of hot or even warm water into a propane cylinder can cause a rapid rise in tank pressure. This could result in relief valve operation or even a tank quickly reaching 100% capacity. This rise in pressure and or volume may add additional stress on an already compromised container causing a catastrophic failure. Prior to connecting the water line to the injection kit flow enough water to clear ALL warm water from the lines and the pump.

In incidents where the propane temperature is less than 32 degrees F the injected water will increase the pressure in the tank. But, as the propane in the tank cools the water, it may turn to slush and slow the leak until enough ice accumulates plugging the leak. For this to occur, water must be injected in sufficient volume to allow it to freeze prior to additional water being added that would warm the already cold water. If water is continuously injected a vapor pressure is not being flared, the propane tanks pressure will continue to rise and the warmed water may not create an ice plug. Attaching a flare capable of exceeding the tank's vapor production capability (1-inch flare recommended) and flaring vapor should assist with the creation of an ice plug (Figure 4.156).

Potential Vapor Cloud

Propane has an expansion ratio of 270 to 1 which means that for every gallon of propane liquid that escapes the container, it expands and creates 270 gallons of vapor. But it doesn't stop there, the 270 gallons of vapor exists at 100% concentration (too rich to burn). Propane has a flammable range between approximately 2% to 10% in air. The accompanying table expands on (pun intended) the relationship between propane's expansion ratio, flammable range and just how large an ignitable vapor cloud can be. Looking at the table you see a standard 500-gallon propane tank holds 400 gallons of liquid propane when filled to 80%. If its entire contents were released the liquid would have the potential of creating more than 5,000,000 gallons of ignitable vapor. Normally managing an uncontrollable liquid leak with water fog is most departments only option to mitigate the resulting vapor cloud. But what if you could convert the leak to water before the release created such a large cloud? With water injection you could possibly convert it to a water leak after only loosing "X #" of gallons? How would you manage an uncontrollable liquid leak if it happens today?

Where to Inject Water?

Access into and out of propane tanks can be challenging depending on the vessel's connection options. Cylinders with multi-valves may have one connection marked as liquid or fill and provide a designated location to inject water through. Or the tank may only have one valve where its filled and product removed limiting your options. Some tanks and cylinders have a separate connection known as a "Liquid Withdrawal Valve". A liquid withdrawal valve can be located on the bottom of the container or

Figure 4.156 LPG vapor being flared off to reduce the pressure inside the tank so that water can be injected.

mounted on top as part of a multi-valve or be a completely separate unit with piping to reach the bottom of the tank known as a dip tube.

Transports, bobtails and bulk storage tanks typically have multiple connection points, some are liquid or vapor IN, some are liquid or vapor OUT. Larger systems typically have a connection labeled as "Liquid" or "Spray Fill", it's used to fill the tank and typically have a check valve in the line.

Check valves allow product flow in only one direction. Tanks with a designated "Vapor" connection allow a flare to remain connected and pressure monitored during the entire water injection process. Vapor out connections normally contain an excess flow valve that allows product movement in two directions. Water can be injected through and vapor pressure removed through an excess flow valves if need be.

Special attention must be paid to any connection that does not have an ACME or other typical hose connection. Never remove a plug with NPT (National Pipe Threads) from a tank under pressure. The third valve from the left on a tank utilizes a quick connect coupling and requires a special fitting to gain access.

Smaller tanks usually contain a single valve and are limited to only one access point for pressure readings, flaring, and water injection. The tactics required for a single valve tank are considerably different than the storage tanks discussed above.

In the picture on the right the Fire Department was dispatched to a leaking Bobtail with an uncontrollable liquid leak. Fortunately for them, the truck was in a desolate area and they were able to let it bleed off without incident (http://www.bemidjipioneer.com/content/road-closed-after-propane-leak).

But what happens when it occurs in town (http://www.tbo.com/central-tampa/tampa-firefighters-propane-truck-leak-under-control-18336)? Propane's vapor density is 1.5 and air is 1, any released propane vapor will sink and find the lowest areas such as basements, sewers, low areas or travel along with a moving body of water. Propane vapors moving with flowing water can travel miles.

So, let's back and look at the two incidents above:

Incident #1, motor vehicle accident involving a bobtail. As you arrived on scene you had identified a passenger car wedged under the bobtail and the front passengers were trapped and need extricated. The accident had damaged the bobtail causing a liquid leak that's producing a large vapor cloud exposing the trapped victims and responders to the very real possibility of a flash fire. You've now discovered that the truck's normal flow control options are nonfunctional and cannot be used to stop the liquid leak.

Tactical Option #1, using fog streams continuously flow water pushing air and water towards the leaking area and start the extrication process and hopefully the leaking propane does not ignite.

Tactical Option #2, Identify that the bobtail is a candidate for water injection (on its wheels, leaking liquid, etc.). Following the kits instructions and your training, connect your water injection kit. Inject water to the truck and flow enough water to convert the propane leak to a water leak. At this point the accumulated vapors will begin to dissipate allowing for a safer rescue. A water fog could be used to speed up the dissipation of vapors. As soon as possible connect a high-volume flare (1 inch) and start reducing the tanks internal pressure and temperature. Manage the water injection process to match the water leak and continue to flare vapor. If the amount of water being injected has sufficient weight to compress the bobtail's springs endangering the occupants, use cribbing to support the frame.

Incident #2, 100-pound tank leaking through the bottom. As you arrive on scene the food vender says he just had the tanks filled and both valves are open. You know that this type of tank normally does not have a lower connection option so it is believed that the tank has probably rusted through the tank shell. Any tank that's shell is compromised requires immediate actions.

Tactical Option #1: Due to the tank shell being compromised you do nothing except evacuate the area and wait. 6 hours later both tanks have emptied their contents and the vapor has dissipated.

Tactical Option #2: Due to the belief that the tank shell is compromised, first arriving units decided to evacuate the immediate area. Command has decided to use water injection to convert the liquid propane leak to a water leak but not until the tanks internal pressure has been reduced. Under the protection of a fog stream (pushing and dissipating product vapors) the entry team shuts off the non-involved tank. Then using the appropriate connectors, they connect to the leaking tank and start flaring operations to reduce the internal pressure and lessen the potential of a catastrophic failure. Flaring continues at as high a rate as possible until the pressure is visibly lower (reduced flare flame due to auto-refrigeration). The tanks volume is identified using a Thermal Imaging Camera (TIC) and a frost line is visible. The tank appears to be approximately 60% full and is a candidate for water injection. The entry team connects the water injection kit, identifies the pressure using a gauge assembly and injects cool water. Within seconds the leak converts to a water leak and the vapors start dissipating. Water is injected until the tank reaches 80% of its capacity (using the spit gauge) and the pressure is continually controlled and monitored. The entry team reconnects the flare and continues to flare vapor reducing the pressure further and refrigerating the tank. Using auto-refrigeration, it may be possible to freeze the water and stop the leak completely. Once frozen the tank may be able to be moved to a safer location to have the remaining propane vented away. It must be noted that expanding ice has the possibility of causing further damage to the cylinder.

There are 0.236 gallons per pound of propane. A 100lb tank contains 23.6 gallons.

So how much water will it take? First you need to know what the capacity of the tank is, some are identified by the weight of the propane that it holds while other are identified by its water capacity (WC on the tag).

Most departments and hazmat teams don't have the tools they need to perform water injection operations. Why is that? The most common answer is "we've never needed a water injection kit before." Are you sure, knowing what you know now, have there ever been incidents that water injection could have been used to lessen the hazard and shorten on scene time allowing you to return the area to near normalcy sooner? Just about every HazMat team has a chlorine kit and have never used it for an actual response, but they own it just in case. If you look at a commodity flow study for your jurisdiction which do you think you have more of, chlorine or propane? Even cities that don't have residential propane tanks have the potential for motor vehicle accidents involving bulk transports or alternative fuel vehicles. It might be time to add a new weapon to your arsenal.

Conclusion

When Ludwig Benner worked at the National Transportation Safety Board (NTSB) they did an accident risk study involving hazardous materials and concluded that emergency responders had a 10,000 times greater chance of death or injury than anyone else. This is one of the things that prompted him to develop decision models to help responders make better decisions when responding to hazardous materials.

> ***Author's Note:*** *When I spoke to him on the phone the other day he said: "the greatest challenge we face today in hazardous materials emergency response is to inform a new generation with the knowledge to safely work with those materials." One of the main reasons that prompted me to write Hazmatology: The Science of Hazardous Materials was much the same motivation as Ludwig Benner almost 50 years ago.*

Bibliography

Volume 4

Association of American Railroads, https://www.aar.org/.

ATF, Bomb Threat Checklist, https://www.atf.gov/file/104816/download.

BBC News, Lac-Megantic: The Runaway Train That Destroyed a Town, Lac-Megantic, Quebec, CA July 6, 2013, Train Derailment Crude Oil Fire, https://www.bbc.com/news/world-us-canada-42548824.

Benner, Ludwig, DECIDE and GEBMO Incident Management Models, 1970s.

Big Spring Daily Herald, Texas, July 30, 1956.

Burke, Robert, Biological Terrorist Agents: Part 2 – Viral Agents, *Firehouse Magazine*, July 1, 2002.

Burke, Robert, Chemistry of Clandestine Methamphetamine Drug Labs, *Firehouse Magazine*, December 31, 2003.

Burke, Robert, *Counter Terrorism for Emergency Responders*, Third Edition, CRC Press, Taylor & Francis Group, 2018.

Burke, Robert, Cryogenic Liquids, *Firehouse Magazine*, November 30, 1996.

Burke, Robert, Decontamination for Hazmat & Terrorism, *Firehouse Magazine*, February 1, 2018.

Burke, Robert, Decontamination Options for Hazmat & Terrorist Incidents, *Firehouse Magazine*, December 31, 2000.

Burke, Robert, Emergency and Technical Decontamination for Hazmat and Terrorism, *Firehouse Magazine*, June 1, 2005.

Burke, Robert, Emergency Response to Agricultural Ammonia Releases, *Firehouse Magazine*, July 1, 2014.

Burke, Robert, Ethanol – Part 2: The Manufacturing Facility, *Firehouse Magazine*, March 3, 2010.

Burke, Robert, Ethanol – Part 3: Transit and Fixed Facilities, *Firehouse Magazine*, May 5, 2010.

Burke, Robert, Handling Anhydrous Ammonia Emergencies, *Firehouse Magazine*, February 28, 2002.

Burke, Robert, *Hazardous Materials Chemistry for Emergency Responders*, Third Edition, CRC Press, Taylor & Francis Group.

Burke, Robert, Hazardous Materials Dangers in Confined Spaces, *Firehouse Magazine*, May 1, 2004.

Burke, Robert, Hazmat Response in the Heartland, *Firehouse Magazine*, November 30, 2008.

Burke, Robert, Hazmat Studies: Fire and Police Combine For Hazmat Response: Part 2, *Firehouse Magazine*, September 1, 2013.

Burke, Robert, *Hazmatology: The Science of Hazardous Materials*, Volume One, CRC Press, Taylor & Francis Group, 2020.

Burke, Robert, *Hazmatology: The Science of Hazardous Materials*, Volume Two, CRC Press, Taylor & Francis Group, 2020.

Burke, Robert, *Hazmatology: The Science of Hazardous Materials*, Volume Three, CRC Press, Taylor & Francis Group, 2020.

Burke, Robert, How Picatinny Arsenal's Hazmat Team Utilizes Its Hazmat Robot, *Firehouse Magazine*, August 1, 2007.

Burke, Robert, Inside the Greater Cincinnati Hazmat Unit, *Firehouse Magazine*, May 1, 2017.

Burke, Robert, Inside the Houston Hazmat Team, *Firehouse Magazine*, February 1, 2019. Visited Houston Hazmat Team, road with them for a day, gathered information about incidents from Houston Hazmat Team members.

Burke, Robert, Lessons Learned from Anhydrous Ammonia Incident, *Firehouse Magazine*, April 1, 2017.

Burke, Robert, Orlando Fire Department Hazmat Response, *Firehouse Magazine*, July 1, 2006.

Burke, Robert, Part 1: Ethanol-Physical and Chemical Aspects, *Firehouse Magazine*, December 29, 2009.

Burke, Robert, Plastics & Polymerization: What Firefighters Need To Know, *Firehouse Magazine*, February 28, 1999.

Burke, Robert, Preparing Dispatchers for Hazmat & Terrorist Emergencies, *Firehouse Magazine*, September 1, 2001.

Burke, Robert, Preparing the Hospital for Victims of Hazmat & Terrorist Incidents, *Firehouse Magazine*, October 31, 1999.

Burke, Robert, Preventing Chemical Emergencies in Schools, *Firehouse Magazine*, July 1987.

Burke, Robert, Protecting the Pentagon, *Firehouse Magazine*, May 2, 2011.

Burke, Robert, Rescuing the Rescuer: Allegheny County's Specialized Intervention Team, *Firehouse Magazine*, November 30, 2006.

Burke, Robert, Safe Response to Aerial Crop-Spraying Accidents, *Firehouse Magazine*, February 1, 2012.

Burke, Robert, Tactical Response to Incidents Involving Crude Oil, *Firehouse, Magazine*, September 1, 2015.

Burke, Robert, The Dangers of Synthetic Opioids, *Firehouse, Magazine*, September 1, 2017.

Burke, Robert, The Phenomenon of Spontaneous Combustion, *Firehouse Magazine*, October 31, 2003.

Burke, Robert, Understanding Ammonium Nitrate, *Firehouse Magazine*, December 1, 2013.

Burke, Robert, Understanding Chlorine, *Firehouse Magazine*, February 29, 2004.

CDC, National Institute for Occupational Safety and Health (NIOSH). https://search.cdc.gov/search/?query=Biological+Protective+Equipment&sitelimit=NIOSH&utf8=%E2%9C%93&affiliate=cdc-main.

CDC, National Institutes for Health, Chemical Segregation Storage Table, https://www.ors.od.nih.gov/sr/dohs/Documents/General_Chemical_Storage_Compatibility_Chart.pdf.

Chemical Safety Board (CSB), CSB Releases Call to Action on Combustible Dust Hazards, https://www.csb.gov/csb-releases-call-to-action-on-combustible-dust-hazards/.

Chemical Safety Board (CSB), DPC Enterprises Festus Chlorine Release, Festus, MO, August 14, 2002, https://www.csb.gov/dpc-enterprises-festus-chlorine-release/

Chemical Safety Board (CSB), DPC Enterprises Glendale Chlorine Release, Glendale, AZ, November 17, 2003, https://www.csb.gov/dpc-enterprises-glendale-chlorine-release/.

Chemical Safety Board (CSB), West Fertilizer Explosion and Fire, West Texas, TX, January 28, 2016, https://www.csb.gov/west-fertilizer-explosion-and-fire-/.

CHEMTREC, https://www.chemtrec.com/. Visits to facility in Washington D.C. and telephone interview with director.

Common Sense Decontamination, Allegany County, PA Green Team.

Corpus Christi, TX, Fire Department, I visited the department and spent time with the AERO team and went on a mission with them to survey flooding west of the city.

Corpus Christi, TX, Fire Department's Drone Program Makes History, October 18, 2016, http://news.cctexas.com/news/fire-department-s-drone-program-makes-history.

DHHS, Chemical Hazards Emergency Medical Management, Decontamination, https://chemm.nlm.nih.gov/decontamination.htm.

DOT, PHMSA, https://www.phmsa.dot.gov/.

EPA, Chemical Accident Investigation Report, Terra Industries, Inc. Nitrogen Fertilizer Facility Port Neal, Iowa, December 13, 1994.

Ethanol Incidents - Ethanol and the Fire Service-LibGuides at University of Illinois. http://guides.library.illinois.edu.

FAA, Unmanned Aircraft Systems (UAS), https://www.faa.gov/uas/.

FBI, Hazardous Materials Response Unit, Dry Decontamination, https://www.fbi.gov/.

FEMA, Center for Domestic Preparedness, Anniston, AL, https://cdp.dhs.gov/.

FEMA, National Fire Academy, Chemistry for Emergency Response Instructor Guide, 2017.

FEMA, National Fire Academy, Hazardous Materials Incident Management, Instructor Guide, 2014.

FEMA, National Fire Academy, *Initial Response to Hazardous Materials Incidents: Basic Concepts, Student Manual*, July 2003.

FEMA, National Fire Academy, *Initial Response to Hazardous Materials Incidents: Concept Implementation, Student Manual*, July 2003.

FEMA, National Fire Academy, *Operating Site Practices, Student Manual*, 2014.

Fire Ox, Firefighting Robotic Vehicle System, https://www.fireapparatusmagazine.com/2014/08/22/firefighting-robotic-vehicle-system/.

Freight Train 292-16 and Subsequent Release of Anhydrous Ammonia Near Minot, North Dakota January 18, 2002.

Hanover-Adams, *The Evening Sun*, Thursday May 26, 1994.

Huffman, Ronald D., Emergency Water Injection -101, Lieutenant/EMT/Hazmat Tech at New Castle, IN, Fire Department, Owner Responder Training Enterprises, LLC. https://www.respondertraining.com/emergency-water-injection-kit/.

International Association of Firefighters (IAFF), https://www.iaff.org/.

International Maritime Organization (IMO), Safety and Security of Shipping, http://www.imo.org/en/OurWork/Pages/Home.aspx.

Lesak, David, GEDAPER Incident Management System, 1980s.

Lowell Sun, Massachusetts, June 24, 1949.

Milwaukee Fire Historical Society, Jim Ley, Milwaukee Fire Department.

Mississauga, Ontario, Canada, November 10, 1979, Derailment Fires Explosions. Mississauga Fire Department telephone interviews and email correspondence.

National Archives, Sciences, Engineering and Medicine, TRID Data Base, Highway Accident Report: Propane Truck Collision with Bridge Column and Fire, White Plains, New York, July 27, 1994, NTSB, https://trid.trb.org/view.aspx?id=459489.

National Institutes for Occupational Safety and Health (NIOSH), https://www.cdc.gov/niosh/index.htm.

Nebraska State Fire Marshal, Investigation Report, Midwest Farmers COOP, Tecumseh, NE, Anhydrous Ammonia Release, March 20, 2014.

NFPA 704: Standard System for the Identification of the Hazards of Materials for Emergency Response. https://www.nfpa.org/codes-and-standards/all-codes-and-standards/list-of-codes-and-standards/detail?code=704.

NFPA, *Fire Protection Handbook*, 15th Edition, Section 4, Chapter 5, Gases. National Fire Protection Association.

NIOSH, Firefighter Fatality Investigation, One Fire Fighter Killed and Eight Fire Fighters Injured in a Dumpster Explosion at a Foundry Wisconsin, https://www.cdc.gov/niosh/fire/reports/face200931.html.

NTSB Investigation Report, Derailment of Canadian Pacific Railway. https://www.ntsb.gov/news/events/Pages/Derailment_of_a_Canadian_Pacific_Railway_Train_near_Minot_North_Dakota_on_January_18_2002.aspx#:~:text=The%20National%20Transportation%20Safety%20Board%20determines%20that%20the,the%20breaking%20of%20the%20rail%20at%20the%20joint.

NTSB Investigation Report, Derailment with Hazardous Materials Release, Shepherdsville KY January 16, 2007, Crude Oil Train Derailment & Fire, https://www.ntsb.gov/investigations/AccidentReports/Reports/RAR0401.pdf.

NTSB Investigation Report, Propane Truck Collision with Bridge Column and Fire, White Plains, NY, July 27, 1994. https://ntsb.gov/investigations/AccidentReports/Pages/HAR9502.aspx.

OPM.gov, https://www.opm.gov/policy-data-oversight/performance-management/teams/building-a-collaborative-team-environment/.

OSHA Standard 1910.120, https://www.osha.gov/laws-regs/regulations/standardnumber/1910/1910.120.

Philadelphia Fire Department, Hazardous Materials Team, https://www.phila.gov/departments/philadelphia-fire-department/.

Philadelphia Fire Department, Visited and interviewed firefighters with the help of Battalion Chief Bill Doty and Hazmat Administrative Services Chief Mike Roeshman on all events that appear in this publication in the City of Philadelphia. Further assistance provided by Harry McGee at the Philadelphia Fire Museum.

Shreveport Public Library. The Times Shreveport-Bossier, September 18, 1984.

Thermite RS1-T4, World's First Purpose Built Robotic Fire Fighting Solution, http://www.roboticfirefighters.com/.
Thor: Tactical Hazardous Operations Robot, http://www.navaldrones.com/SAFFiR.html.
Turbine Aided Firefighting Machine (TAF 20), https://www.alphr.com/the-future/1002221/meet-taf20-the-turbine-aided-firefighting-robot-of-the-future.
U.S. Army, Edgewood Chemical and Biological Command (ECBC), Dry Decontamination, https://www.cbc.ccdc.army.mil/.
United Nations, *Globally Harmonized System of Classification and Labelling of Chemicals (GHS)*, Second Revised Edition. United Nations Geneva Switzerland.
Utah Valley University, Emergency Services, Jack Rabbit Project, April 10, 2010, https://www.uvu.edu/es/jack-rabbit/.
Webster's Dictionary, http://www.webster-dictionary.net/
West Fire Department, St. Louis County, MO, Drone Operations. I visited the West Fire Department and witnessed a drone demonstration.

Index